CW01640925

Defect Avoidance Manual

New Build

Prepared by Construction Audit Ltd
for

HAPM

HOUSING ASSOCIATION PROPERTY MUTUAL LIMITED

Foreword

This publication will greatly help to provide better housing by minimising defects whether from design or construction. Although written primarily with housing associations in mind, it will be of value in both the public and private sector. The practical approach to defect prevention, together with the clear and orderly presentation will help all those who provide and care about housing.

The manual was commissioned by Housing Association Property Mutual Limited (HAPM), a mutual insurance company set up in 1990 by a group of housing associations in response to Government changes to the funding and subsidy arrangements in respect of housing association activity.

As a mutual scheme, HAPM is controlled by and exists solely for the benefit of its member housing associations who comprise a substantial proportion of those engaged in development. Together, they are able to tailor the insurance cover specifically to match their needs.

HAPM itself breaks new ground in defect prevention and insurance for houses, flats and hostels whether newly built or rehabilitated. Its unique policy gives cover for 35 years against non-structural as well as structural defects, and a wide range of incidental losses.

An integral and innovative part of HAPM's structure is its sophisticated approach to defect identification and prevention. The principles have been encapsulated in this manual to help identify sensible methods of avoiding common building defects. Pragmatic advice on current standards and accepted good practice adds to the benefits of the manual.

I very much hope that all concerned with the planning, development and construction of housing will find this manual of assistance in producing better homes for the occupiers with fewer maintenance and management problems.

Nicholas Abbott, FRICS
Chair of the Board of Directors of HAPM

Introduction

This manual is produced by Construction Audit Limited (a subsidiary of defect specialists NBA Building Performance Services) in their role as technical auditors for the Housing Association Property Mutual Limited.

The purpose of the technical audit is to identify and eliminate features of a scheme which may give rise to a claim under the policy. Minimisation of risk will benefit all members of HAPM by enabling premiums to be kept as low as possible.

The basis of the audit system is the extensive collective knowledge of common defects of the Technical Audit Unit's experienced professional staff. This manual lists common defects together with appropriate guidance and advice on how they may be avoided.

It is important to stress that HAPM is not setting new standards but merely seeking to ensure that good practice as already defined is followed.

The audit process is divided into two parts. The first is a document and drawing appraisal and the second is one or more site visits by a Technical Audit Unit Inspector. After the drawing and document appraisal, any matters of concern will be notified to the housing association who will be responsible for taking the appropriate action. The site visit will enable the Inspector to re-assess the design appraisal in the light of what is found on site and will also enable him to form an opinion on the general quality of workmanship.

The paragraph headed HAPM GUIDANCE is intended to advise associations of the information and documentation they need to submit for audit and what items will be checked.

The approach being adopted by the Technical Audit Unit is new in the UK housing development field and the Technical Audit Unit welcomes any comments and feedback from HAPM members and others.

The continuing collation and feedback of technical data by the Technical Audit Unit will be a major part of its future role. In time, this will result in a resource of knowledge of considerable magnitude for the benefit of all members and all those concerned about housing. Discussions are in hand to ensure that this information is published and available to all those who wish to provide "better" housing.

HAPM would like to thank all member housing associations who have supplied photographs.

Now also
HAMP COMPONENT LIFE MANUAL
ISBN 0419 18360 4
Published SPON

Further copies of this publication can be obtained, quoting reference EP7, from:

BRE Bookshop
Building Research Establishment
Garston, Watford WD2 7JR
Telephone: 0923 664444

This manual was commissioned by Housing Association Property Mutual Limited and written by Construction Audit Limited. It is published and distributed by the Building Research Establishment. All technical enquiries should be addressed to Housing Association Property Mutual Limited, International House, 26 Creechurch Lane, London, EC3A 5BA.

First published 1991

ISBN 0 85125 503 5

Designed and produced by NBA Tectonics

Contents

Each of the seven sections of the Manual contains advice on a number of related elements of a typical building.

Within each section the guidance on each element is divided into the same four topic headings:

- Design
- Details
- Materials
- Workmanship

It is important to realise that the Manual does not offer a comprehensive guide to building design and construction but only covers those building elements which have been found to be problematic and likely to give rise to defects and claims under the HAPM policy.

HAPM
HOUSING ASSOCIATION PROPERTY MUTUAL LIMITED

1
FOUNDATIONS
GROUND FLOORS

2
STEEL OR CONCRETE FRAMES
EXTERNAL MASONRY WALLS
EXTERNAL TIMBER FRAMED WALLS
MASONRY SEPARATING WALLS
TIMBER FRAMED SEPARATING WALLS

3
INTERMEDIATE FLOORS AND STAIRCASES
PARTITIONS

4
PITCHED ROOFS
FLAT ROOFS
DORMERS
CHIMNEYS AND FLUES

5
WALL CLADDINGS
PURPOSE -MADE JOINERY
BAYS, PORCHES, CANOPIES, ORIELS
EXTERNAL BALCONIES, STAIRS, RAMPS

6
ELECTRICAL INSTALLATION
HOT AND COLD WATER INSTALLATIONS
GAS HEATING
ABOVE GROUND DRAINAGE
SHAFTS, DUCTS, CHUTES

7
BELOW GROUND DRAINAGE
EXTERNAL WORKS
OUTBUILDINGS

FOUNDATIONS

GROUND FLOORS

1

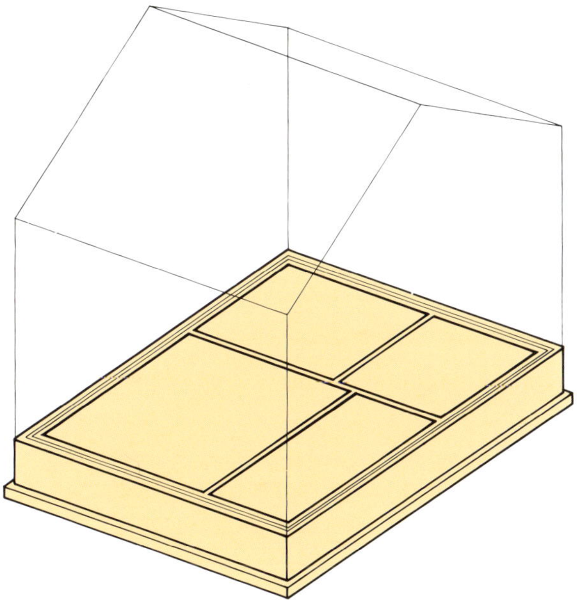

HAPM

HOUSING ASSOCIATION PROPERTY MUTUAL LIMITED

Design

COMMON DEFECTS	CRITERIA/GUIDANCE
Site conditions not fully investigated, and hazardous ground conditions requiring special consideration not recognised	A soil survey consisting of trial holes, borings and chemical analysis should be carried out by a specialist firm before any detailed design is started. Areas with high water tables, liable to flooding or having pockets of made ground (fill) or chemical contamination should be especially noted. Refer to BS 5930: 1981 *Code of practice for site investigations*.
Footings not taken to adequate bearing	The results of the site investigation should be used to determine an appropriate foundation design. Where dwellings are constructed on made ground or on landfill sites, special precautions should be taken to ensure that footings are taken down to soil of adequate bearing capacity. Vegetable soil, rubbish, old buildings, tree roots should be removed.
Footings not taken to sufficient depth	Footings should be taken below tree root level. In the case of dwellings near large trees, this could result in deep footings – possibly 5 metres – when using piled foundations and ground beams would be more economical. The advice of a Structural Engineer should be sought for the design of any foundation deeper than 2.5m. The minimum depth of a footing in shrinkable clay should be one metre.
Inadequate 'flexibility' incorporated in foundation design	Where different types of foundation are used, for instance piles and ground beams for one end of a building and strip footings at the other end, or where measurable changes occur to the bearing capacity of ground under a foundation, movement joints should be incorporated to accommodate any differential settlement. Movement joints should be extended vertically through the superstructure.
Inadequate attention to adjacent buildings	Footings should be designed so that: • they do not apply any surcharge to foundations of adjacent buildings • they are not surcharged by the adjacent foundations.
Settlement damage due to mine workings	In mining areas, records should be checked for any evidence of old mine workings, and mining authorities should be contacted for details of future developments. Design proposals should be submitted to the mining authority for approval.
Movement suffered by dwellings built on slopes	In clayey areas on slopes steeper than 10%, ground is liable to slip or creep. Footings should always be horizontal, with steps as necessary, to accommodate the slope of the ground. In these circumstances, trench fill foundations can also act as retaining walls, provided they are so designed by a Structural Engineer.

FOUNDATIONS

Design

COMMON DEFECTS	CRITERIA/GUIDANCE
Under-estimation of effects of single trees, rows of trees or removal of scrub and hedges	The potential effects of trees, hedges and shrubs (whether retained, removed or planned) should be carefully evaluated in relation to foundation design.

HAPM GUIDANCE

A soils investigation and report will be expected for all developments. Foundations should comply with *Approved Document A1/2* of the *Building Regulations*. The information given in BS 8004: 1986 *Code of practice for foundations* is largely superseded by Section 6 in BS 8103, Part 1: 1986 *Structural design of low-rise buildings: Code of practice for stability, site investigation, foundations and ground floor slabs for housing* where a simple guide is given to choice of foundations. Refer also to BRE *Digest 240, 241, 242* and *298*, BRE *Defect Action Sheet 96* and *The Effects of Trees on Foundations in Clay Sub-soils* published by the Seminar for Structural Engineers of London Boroughs (SEMSELB). Whenever the simple design rules are not applicable the design should be prepared by a Structural Engineer.

FOUNDATIONS
Details

COMMON DEFECTS	CRITERIA/GUIDANCE
No provision for clay heave	Where piles and ground beams have been used in shrinkable clays, heave may be accommodated by using compressible material at the sides and below the soffit of ground beams.
Edge of footings liable to become unstable	On sloping sites (especially at the lowest building at the edge of a site) care should be taken that the footing never becomes unstable due to the erosion of surrounding ground. It is good practice to provide a concreted path or driveway to maintain the ground surface.
Cold bridging at edges of foundations and thresholds	Cold bridging should be avoided by careful detailing. Refer to BRE *Thermal insulation: avoiding risks* which gives guidance for all common situations.
Wall cavities not solidly filled below dpc level	Walls below the level of the dpc need to be sufficiently robust to retain fill and withstand fill compaction. If any cavity is solidly filled to just below the level of the dpc, the wall can act more efficiently and traps for moisture at the bottom of the cavity can be avoided. However, sufficient margin should be allowed between the level of the dpc and the top of any solid fill for the inevitable accumulation of bricklaying mortar (see Topic 2.2 for more guidance).
Inadequate waterproofing at steps	Care should be taken in detailing at steps to ensure that continuity is maintained between the dpcs in walls and the dpms in floor slabs.
Wall unstable between footing and ground floor	Attention should be paid to the provision of wall ties between the footing and ground floor as much as for walls in other locations. When the filling material is placed against a wall, care should be taken that the fill is brought up equally on each side of the wall. When the filling material is to be higher on one side of a wall, refer to BS 8103: Part 1: 1986 *Structural design of low-rise buildings: Code of practice for stability, site investigation, foundations and ground floor slabs for housing.* When the difference in level of fill is more than four times the wall thickness, the wall should be designed by a Structural Engineer.

HAPM GUIDANCE

Details showing typical longitudinal and cross-sections through foundations should be provided. Arrangements at steps should be detailed showing proposed waterproofing measures. The method of constructing any movement joints should be indicated clearly.

Materials

COMMON DEFECTS	CRITERIA/GUIDANCE
Insufficient resistance of specified materials to sulphate attack	Where sulphates are present in the ground, consideration should be given to using sulphate-resisting cement for the concrete and mortar. Allowance should be made for additional cover to reinforcement. Bricks and blocks which are resistant to chemical attack should be used below the level of the dpc.
Incorrect specification for concrete	Concrete mixes should be specified in accordance with BS 5328: 1981 *Methods for specifying concrete, including ready-mixed concrete* and have a minimum cement content of 220kg/m^3.
Aggregates for concrete liable to chemical change and reductions in strength	Care is required in the selection of aggregates for concrete in the light of evidence of failure from ASR (alkali-silica reaction) and the use of iron pyrites-rich crushed granite (Mundic). There is no British Standard to assist in the selection of aggregates.
Board not sufficiently compressible	In situations where there is a risk of clay heave, a compressible board may be specified to absorb ground expansion. Special proprietary products that have been designed for this purpose should be selected. Note that standard polystyrene and similar products may not be sufficiently compressible.

HAPM GUIDANCE

All materials for foundations and structure below the level of the dpc should conform to the requirements of the relevant British Standards. Types of brick, block, cement and mix to be employed should be selected to ensure satisfactory service for the life of the building. Factors to be taken into account include:
(a) the nature of the ground and the possibility of chemical attack therefrom
(b) the physical and chemical characteristics of the materials themselves
(c) the quality of workmanship that may be expected
(d) the degree of supervision to be provided.

If ready-mixed concrete is to be used it should be supplied by a company certified by the Quality Scheme for Ready Mixed Concrete (QSRMC). If site batching is envisaged, weigh-batch plant should be used and suitable quality assurance procedures specified.

Workmanship

COMMON DEFECTS	CRITERIA/GUIDANCE
Errors in setting out	Check dimensions should always be taken to all existing landmarks and benchmarks. The *Manual of Setting-out Procedure* published by the Construction Industry Research and Information Association (CIRIA) gives good guidance.
Obstructions not formally reported and instructed	Any obstructions found during excavation should be reported to the Supervisor. Such obstructions could include: • old building foundations • basements or cellars • tanks • wells • waste deposits • existing drains • services, etc.
Collapse of excavations	Since mixer trucks may impose additional loads on trench sides, adequate support for the sides of excavations should be provided.
Soft spots in excavations	Soft spots should be dug out and the bottom of all footings should be approved by the Supervisor before any concrete is poured. Especially in limestone areas, the possibility of swallow holes should be investigated. The last 50mm of any excavation should not be completed until immediately before pouring the concrete.
Inadequate liaison for special foundations	When a Structural Engineer has been consulted for the design of any foundation, he should visit the site during construction to ensure that the designed foundation is suitable for the ground conditions encountered.
Inadequate protection of new concrete	Newly-poured concrete should be protected: • in hot weather against too rapid drying out to prevent shrinkage cracks • in cold weather against frost damage and freezing.
Inadequate batching of site-mixed concrete	The proportions of materials for any structural concrete should never be gauged (measured) by the shovelful or bucketful. The proportions should be weigh-batched. If weigh-batching equipment is not available, gauge boxes should be used. The weighing equipment should be checked at regular intervals.
Inadequate attention paid to making test cubes	Each day test cubes, usually in sets of three, should be made for each item of work. The test cubes should be made strictly in accordance with the requirements of BS 1881: Part 101: 1983 *Testing concrete: Method of sampling fresh concrete on site* by a reasonably skilled operative. The preferred sizes are: • 150mm cubes for all concrete work • 100mm cubes for testing mortar strength.
Inadequate or excessive compaction of concrete footings	The concrete in footings should always be compacted in place by means of rodding, tamping or vibrating to achieve a dense, durable mix and to remove all air bubbles. Care should be taken not to overvibrate the concrete which can cause the mix to segregate leaving weak spots in the footing.

FOUNDATIONS

Workmanship

COMMON DEFECTS	CRITERIA/GUIDANCE
Footings disturbed by following trades	When concrete footings have been poured, they should be left for at least three days before operations by subsequent trades are allowed to proceed so that the concrete may acquire sufficient strength. This period may need to be increased in cold weather when concrete takes longer to acquire its strength. Similarly, plant should not be allowed to cross over newly-poured concrete for the same period.
Shuttering for ground beams not strong enough	Ground beams are often cast above the excavated level using timber shuttering. If the timber shuttering is not strong enough, it may deflect or partially collapse allowing the concrete to settle and expose or reduce the cover to the reinforcement steel.

HAPM GUIDANCE

Site inspections by Supervisors are necessary during construction of foundations in conjunction with the local authority's Building Inspector to ensure that adequate bearing is being provided. Section 11 in BS 8004: 1986 *Code of practice for foundations* gives guidance on safety precautions during construction.

GROUND FLOORS

Design

COMMON DEFECTS	CRITERIA/GUIDANCE
Settlement of slab where concrete ground-supported slabs placed on fill greater than 600mm deep	Fill material should be compacted using mechanical compactors in layers not exceeding 225mm of loose depth. For fill greater than 600mm deep, adequate compaction cannot be guaranteed and suspended floors should be used. Deep fill is most likely to be the case on sloping sites.
Excess of service trenches under floor	It is difficult to compact filling material around pipes and cables. Where such services are in trenches more than 600mm deep, they should be treated as a deep fill site.
Moisture penetration to upper surface of a ground-supported slab	The sophistication of damp-proofing measures required will depend on the site conditions and floor finishes used. Refer to CP 102: 1973 *Code of practice for protection of buildings against water from the ground.* Where ground water can exert a hydrostatic pressure, the ground slab will require specialist treatment. Refer to BRE *Defect Action Sheet 35* for the specification of dpcs and dpms.
Vibration nuisance in long span timber suspended ground floors	Traditionally, ground floors spanned between sleeper walls and gave few vibration problems. The current trend towards long span timber suspended ground floors increases the risk of vibration nuisance generated by home laundry and hi-fi equipment and younger pedestrians. Where the use of long span timber suspended ground floors is unavoidable, ensure a robust design by specifying: • blocking between the ends of joists • intermediate blocking • generous joist sizes. Joists supported by hangers are more prone to problems than those built into walls (see Topic 1.2, Details, for more guidance).
Lifting of ground-supported slabs	Lifting (heaving) of ground-supported slabs may result from: • sulphate attack on the concrete • expansive materials being used for filling • expanding clay resulting from the effects of tree or hedge removal. Concrete should be protected from sulphate attack by an adequate dpm. Suspended floors should be used in areas of expansive clays.
Floor layouts changed on site to suit actual position of services	The position of all service pipes, ducts and drains should be drawn and checked before the floor design is finalised.
Failure of raft foundations	Raft foundations may break their backs as a result of settlement or heave if they are not adequately profiled and reinforced. Care should be taken to ensure that any hard fill material is continued under the perimeter downstand to prevent breaking away. The design of raft foundations should always be carried out by a Structural Engineer.
Inadequate protection against radon emissions or methane	On landfill sites where there is a risk of methane, and in Devon, Cornwall and other areas where radon emissions have been identified, a membrane of 1200g polyethylene sheet should be placed over the oversite concrete. This polyethylene sheet should be continuous with the dpcs in adjacent walls and sealed at all laps. In the case of suspended floors, sealing the solum, together with arrangements for efficient natural or mechanical ventilation to the floor void, may also be necessary.

GROUND FLOORS

Design

COMMON DEFECTS	CRITERIA/GUIDANCE
Rot in timber batten floors	In the case of timber floor decks on concrete ground-bearing slabs, a dpc should be placed between the battens and the concrete.
Failure of floor screeds when laid over insulation	It is almost impossible to adequately compact floor screeds over a loose-laid insulating layer. This form of insulation should not be used below screeds. For thermal insulation, a chipboard and insulating foam laminate should be used.

HAPM GUIDANCE

Defects in ground floors have been a traditional source of major problems. Remedial costs are usually high due to the need to decant tenants. The Technical Audit Unit will consider the suitability of the floor system proposed in conjunction with the soils report and information regarding trees and hedges as existing and to be removed. A 'robust' solution will be required to ensure that future maintenance will be kept to the minimum. A Structural Engineer's design will be required where simple floor span tables are not applicable.

Details

COMMON DEFECTS	CRITERIA/GUIDANCE
Incorrect detailing of dpcs and dpms	To avoid rising damp, dpcs and dpms should be continuous. Laps and seals at joints should be carefully considered. Arrangements at steps and staggers should be fully detailed.
Cold bridging at junctions with external walls	Cold bridging at the junctions with external walls is a particular problem with concrete floors. Local internal insulation should be provided. Refer to BRE *Thermal insulation: avoiding risks* which gives guidance for all common situations.
Cold bridging under door thresholds or full-height frames	Brickwork is often set back under cills thus reducing or even eliminating the cavity. If the cavity is maintained it should be detailed to adequately support any floor screed, without forming a cold bridge.
Voids not vented under suspended floors	All voids under suspended timber floors should be ventilated. The guidance on levels of ventilation varies: • BRE *Defect Action Sheet 73*, quoting CP 102: 1973 *Code of practice for protection of buildings against water from the ground,* gives a requirement of 3,200mm^2 per metre run of external wall • *Approved Document C4, Para 1.10(b),* gives a figure of 3,000mm^2 per metre run of external wall • BS 5250: 1989 *Code of practice for control of condensation in buildings* gives a figure of 1,500mm^2 per metre run of external wall or 500mm^2 per square metre of floor area, whichever is the greatest. Voids under suspended concrete floors need not be ventilated according to *Approved Document C4, Para 1.13,* but if not ventilated, a dpm should be incorporated. No figure for ventilation area, if provided, is given. CP 102: 1973 makes no mention of the need to ventilate under concrete suspended floors and BS 5250: 1989 recommends ventilation but gives no figure. Chapter 5.2 in *NHBC Standards* suggests a figure of 600mm^2 per metre run of wall, except for Scotland where the figure is 1,500mm^2 per metre. In ventilating voids under either timber or concrete suspended floors care should be taken to ensure that: • ventilators are placed to allow cross-ventilation • sleeper walls are of honeycomb construction • partitions below floor level incorporate provision for through-ventilation • ventilators are positioned to eliminate 'dead areas' in corners • account is taken of the actual free area of ventilating bricks or plastics ventilators.
Water and heating pipes buried in floor screeds	Pipes should not be buried in floor screeds. Incoming water service pipes should only pass vertically through the floor, not run within it. Purpose-made ducts should be provided for heating pipes but ensure the continuity of the dpc and dpm is not compromised.
Damp penetration around vertical service risers	Where there is hydrostatic pressure, suitable protective measures, such as a sleeve around any pipe that perforates a dpm, should be fully detailed.

Details

COMMON DEFECTS	CRITERIA/GUIDANCE
Timber ground floor joists built into inner leaf of external cavity wall	The ends of joists are extremely vulnerable to rot in this situation. The cavity is often bridged at the base of a wall and the joist ends may protrude into the cavity and trap mortar droppings. It is preferable to do one of the following: • use joist hangers, with appropriate restraint • support the joist on a perforated sleeper wall • thicken the inner leaf up to plate level to provide a supporting ledge.

HAPM GUIDANCE

Details should follow the preferred arrangements for thermal insulation shown in BRE *Thermal insulation: avoiding risks.* Details showing the junction of ground floor and external wall (if a suspended floor then both along the span and across the span), ground floor and separating wall (stepped and level) and door threshold arrangement should be provided. Ventilation provisions should be indicated clearly.

Materials

COMMON DEFECTS	CRITERIA/GUIDANCE
Unsuitable fill material used	Only material that can be easily compacted and which is clean, inert and free from sulphates should be used. Generally, hoggin, gravel, crushed rock, quarry waste are all suitable filling materials. Demolition materials, such as brick or concrete, are rarely adequately graded or cleaned and are best prohibited unless strictly controlled. Chalk is suitable only if there are no groundwater problems.
Corrosion of steel brackets, connectors and straps	Joist hangers to BS 6178: Part 1: 1982 *Joist hangers: Specification for joist hangers for building into masonry walls of domestic dwellings* are made from zinc coated steel sheet having a zinc coating of 600g/m^2, or 275g/m^2 plus a paint treatment. Other steel components should have equivalent protection.
Inadequate durability of insulation materials	Insulants should have adequate compressive strength and should be sufficiently robust to resist tearing and handling forces during construction. Materials should either comply with a British Standard or should be backed by a BBA Agrément Certificate covering their use in this situation.
Insulation materials not closed cell type	Open cell type insulants can absorb water and should never be used in ground contact or below the level of the dpc.
Recycled polyethylene sheet used as dpm material	Recycled polyethylene sheet should never be used in dpcs or dpms since it degrades rapidly even in the absence of sunlight. Recycled polyethylene sheet is difficult to spot but is identifiable to some extent by touch (the recycled particles can be felt protruding above the surface giving a slight sandpaper effect) and the product is more opaque than virgin material. The only sure way of proving the product is by chemical analysis. Proprietary products should have a minimum thickness of 0.3mm, and comply with BS 743: 1970 *Specification for materials for damp-proof courses.*
Dry or wet rot attack of timber items	All timber for ground floor plates, joists and blocks should receive a suitable preservative treatment.
Failure of chipboard decking	Sagging and distortion can occur if the incorrect grade of chipboard is used, or if its edges are not correctly fixed down to joists or noggins. Chipboard should be flooring grade Type C4 or C5 to BS 5669: Part 2: 1989 *Particleboard: Specification for wood chipboard*, in conjunction with a sheet 'tanking' finish in kitchen and bathrooms. The correct thickness of board should be used: • 18mm for joist spacings up to 450mm • 22mm for joist spacings up to 600mm. Ringshank nails or screws should always be used for fixing. Refer to BRE *Defect Action Sheet 32.*

Materials

COMMON DEFECTS	CRITERIA/GUIDANCE
Inadequate specification for floor finish adhesives	Refer to BS 5442: Part 1: 1989 *Classification of adhesives for construction: Classification of adhesives for use with flooring materials* for guidance on specifying adhesives for floor coverings.

HAPM GUIDANCE

The materials used in a ground floor should be sufficiently durable to have a life in excess of the Major Defects Insurance Policy.

Workmanship

COMMON DEFECTS	CRITERIA/GUIDANCE
Inadequate removal of topsoil and contaminated ground	It is bad, but common, practice to leave the topsoil in place until the footings have been completed. This often means that topsoil, vegetable matter and contaminated ground are not fully removed. All topsoil and contaminated ground should be removed before other construction commences.
Poorly compacted fill under ground-supported slabs	Excavations for foundations, drains, services and internal loadbearing wall footings may result in little undisturbed ground remaining in the dwelling area, leading to a risk situation similar to a deep fill site. Extreme care should be taken to ensure that compaction is carried out in layers not exceeding 225mm of loose depth.
Integrity of dpms compromised around vertical risers and intricate shapes	Dressmaking-type skills are often required to adequately dress dpms and dpcs to awkward shapes. BRE *Defect Action Sheet 36* gives guidance.
Precast floor units project into cavity	Precast floor units projecting into the cavity can lead to a bridging of the cavity and damp penetration. Ideally, the units should be fabricated from site dimensions but since the resulting delay is generally unacceptable, site staff should be vigilant in ensuring that the dimensions required in the support walls are achieved with the minimum deviation.
Inadequate protection of new concrete surfaces from effects of weather	All new exposed concrete surfaces should be protected for 7 days from the following harmful effects of the weather: • rain • frost • rapid temperature changes • too rapid drying out. Suitable methods of protection include covering the concrete with damp hessian (which should be kept damp), damp sand or proprietary curing membranes. When there is a risk of frost, insulated covers will be required.
Insulating boards or void formers float out of position when concrete is poured	Insulation and void formers should be fixed down firmly before any concrete is poured.
Shrinkage of timber floors	When timber floors shrink it can result in gaps between tongued and grooved boards and boards splitting at fixing points. The moisture content of timber at the time of construction should not exceed 20-22%. In very dry, well-heated buildings, the moisture content may drop to 6-10% and inevitably some shrinkage will occur. Therefore, floor boards should not be laid until the building has been glazed and is watertight. Refer to BRE, Princes Risborough Laboratory *Technical Note No 12.*
Fine blinding sand becomes lost in voids in the material	Blinding will be lost if the fill material is not well-graded and compacted before blinding commences. To avoid the risk of perforating the dpm, a smooth finish is required.

GROUND FLOORS

Workmanship

COMMON DEFECTS	CRITERIA/GUIDANCE
Screeds break up	Particular attention should be given to the preparation of screed materials: • if screeds are too thin or not carefully cured, they will curl and crack • if screed mixes contain too much fine sand or water, shrinkage will cause curling and cracking • if screeds are given too little drying time, construction water will cause problems with most floor surfacing • if screeds are not batched accurately, are not mixed thoroughly without segregation and are not fully compacted, they will break down. Refer to BRE *Defect Action Sheet 51* and *52*.

HAPM GUIDANCE

The Technical Audit Unit Site Inspectors will undertake a 'quality count'. Supervisors should pay particular attention to the preparation of ground and hardcore prior to placing the concrete.

STEEL OR CONCRETE FRAMES

EXTERNAL MASONRY WALLS

EXTERNAL TIMBER FRAMED WALLS

MASONRY SEPARATING WALLS

TIMBER FRAMED SEPARATING WALLS

2

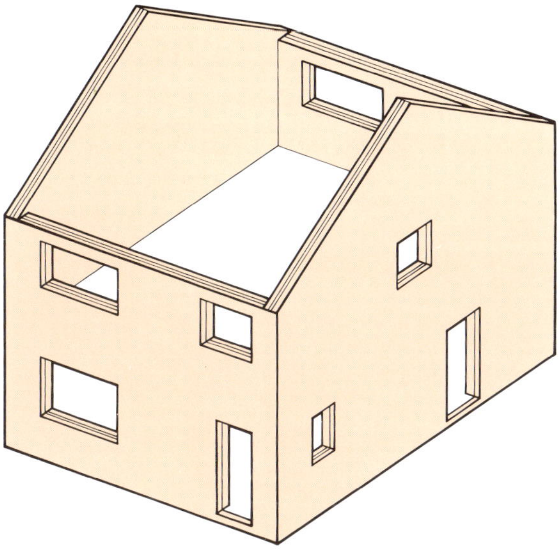

HAPM

HOUSING ASSOCIATION PROPERTY MUTUAL LIMITED

Design

COMMON DEFECTS	CRITERIA/GUIDANCE
Compression failure of infill brickwork	As a practical rule-of-thumb, a reinforced concrete frame shrinks vertically about 2.5mm per storey whereas clay brickwork can expand by about the same amount. Cladding trapped between successive floors may respond by bowing (theoretically about 75-100mm of bow for that amount of shrinkage) or, more commonly, by a mixture of bowing and crushing. Therefore, compression joints and lateral restraint are essential.
Integral concrete nibs too narrow to support brickwork	Supporting brick cladding on integral concrete nibs at floor edges should be avoided since it is difficult to cast the nibs accurately enough to provide sufficient support for the brickwork.
Inadequate protection of steel reinforcement by concrete	Sufficient depth of cover to all steel, including stirrups and tie wires, is essential. In very severe exposure conditions, steel reinforcement may require concrete cover of up to 60mm. It is also important to use low-permeability concrete which is well-graded and compacted and has a low water : cement ratio. Hidden parts of steel reinforcement that could be persistently wet will be protected from both water and carbonation if they are bitumen painted (refer to BRE *Information Paper 12/80*).
Poor specification and quality control of concrete	Refer to BS 5328: 1981 *Methods for specifying concrete, including ready-mixed concrete* and BRE *Digest 326* for guidance.
Inadequate assessment of risk of sulphate attack	When the frame is in contact with the ground, the concrete specification may need to take account of the risk of sulphate attack. See Topic 1.1 for more guidance and refer to BRE *Digest 250*.
Tolerances not fully considered	Manufacturing and erection tolerances should be carefully specified for frames and also for any factory-assembled units that hang on the frame.

HAPM GUIDANCE

The Technical Audit Unit will expect frames to be designed by a qualified Engineer and so full calculations and drawings should be submitted for consideration. Where any element of prefabrication is involved, the specification should indicate manufacturing tolerances to match achievable erection tolerances.

Details

COMMON DEFECTS	CRITERIA/GUIDANCE
Failure of brick slips due to inadequate support of brick cladding	'Brick slip' failure is commonplace along the edges of concrete slabs. BS 5606: 1978 *Code of practice for accuracy in building* advises against carrying brick cladding on integral nibs at the edges of floor or edge beams. Brick panels supported on the floor surface need a soft joint at the top and restraint connections. Guidance should be sought from the British Ceramic Research Association.
No stop ends for cavity trays	When the outer face of columns and the outer edge of floor/edge beams are in same plane, the cavity trays at the level of the floor/edge beams should have upstand ends placed against the columns. Particular care should be taken to achieve a watertight detail.
Condensation due to cold bridging	Exposed frame members are very likely to be cold bridges in an otherwise well-insulated construction unless insulation is applied to the floor, soffit and column faces for a sufficient distance inward from the building face.
Inadequate adjustment provided to cladding fixings	Claddings are expected to be plane and vertical throughout their height. Since deviations in the position of the structural surfaces and cast-in fixings can be relatively large, cladding fixings should be designed to accommodate such deviations.
No provision for movement at heads of non-loadbearing partitions	Shrinkage in concrete columns and deflection in beams and floors can load non-loadbearing partitions. Thermal movements of the roof slab (especially if a fixed 'anchor point' is formed, for example by a lift shaft at one end of the long axis of the building) can result in: • shear cracking at the tops of upper storey partitions parallel to the long axis • rotation of the tops of partitions at right angles to the long axis. Sufficient provision should therefore be allowed for movement at the heads of non-loadbearing partitions.

HAPM GUIDANCE

Details showing junctions of beams, floors and walls and columns and walls should be provided. Waterproofing measures should also be indicated.

Materials

COMMON DEFECTS	CRITERIA/GUIDANCE
Aggregates for concrete liable to chemical change or reductions in strength	Care is required in the selection of aggregates for concrete in the light of evidence of failure from ASR (alkali-silica reaction) and using iron pyrites-rich crushed granite (Mundic). There is no British Standard to assist in the selection of aggregates.
Insufficient protection of structural steelwork against corrosion	The protection system against corrosion should be selected to suit the environment of the steel and whether or not the steelwork is maintainable.
Incorrect specification for steel spacers	Steel reinforcement spacers should be selected to suit: • the level of exposure • the type of shutter to be used • the type of concrete finish required. Bad selection of spacers may result in the provision of a route for moisture or air to reach the reinforcement.
Unapproved or excessive use of concrete admixtures	Care should always be taken to ensure that only approved materials are used and that they are used strictly in the proportions indicated by the manufacturers. Calcium chloride-based additives are now largely prohibited by specifiers.
Use of incorrect concrete mixes	'Prescribed' or 'design' mixes should be specified but note that since prescribed mixes give the specifier control over the amounts of cement, aggregates, etc care should be taken to specify and use them appropriately. Certificates of sampling to BS 1881: Part 101: 1983 *Testing concrete: Method of sampling fresh concrete on site* should be obtained as an important part of quality control.

HAPM GUIDANCE

If ready-mixed concrete is to be used it should be supplied by a company certified by the Quality Scheme for Ready Mixed Concrete (QSRMC). If site batching is envisaged, weigh-batch plant should be used and suitable quality assurance procedures specified.

Workmanship

COMMON DEFECTS	CRITERIA/GUIDANCE
Inadequate compaction of concrete	Segregation, inadequate compaction and excessive water content all reduce the concrete's resistance to carbonation, and hence its durability. Full compaction is essential to prevent damp and carbonation reaching the steel. Vibration should be just enough to achieve mix and compaction. Poker vibrators may be needed where the steel is congested. If segregation is likely or suspected, samples should be taken for immediate comparison with the designed grading. Sampling should be to BS 1881: Part 101: 1983 *Testing concrete: Method of sampling fresh concrete on site*, and tests should be carried out in accordance with the other relevant parts of BS 1881.
Use of incorrect hangers	Steel should not be supported from formwork by any corrodible or porous materials.
Inaccurate formwork	Care should be taken when constructing the formwork since formwork incorrectly sized, positioned or restrained may produce inadequate cover to steel, or a reduction of cover due to the need to hack away concrete that impedes subsequent construction.
Shutters not watertight	The loss of grout from holes and edges of shutters can result in localised reductions in strength.
Inadequate dry-pack to bearings	Dry-pack is a tedious operation and liable to poor workmanship so great care should be taken.
Bad alignment of factory-fabricated components during erection	The margins around factory-fabricated components should be checked before any final tightening, fixing or grouting.

HAPM GUIDANCE

The Technical Audit Unit Site Inspectors will undertake a 'quality count'. All stages of frame construction require skilled supervision and frequent checks.

Design

COMMON DEFECTS	CRITERIA/GUIDANCE
Vertical cracking in masonry	A shortage of movement joints can result in vertical cracking: • at short returns • through parapets • at changes in height, thickness and loading of wall. There is much written on the subject of the provision of movement joints. Some sound and clear guidance is given in BRE *Defect Action Sheet 18* which may be interpreted as a rule-of-thumb and requires a vertical joint: • at every other party wall for clay brick facework • at every party wall for concrete blocks and calcium silicate bricks. Movement joints are also required: • at short returns • at changes in wall height • at changes in wall thickness • where any notching of the facings occurs (ie at position of columns) • to coincide with joints in floor slabs and structural frames. Since parapets are more exposed, less restrained and move much more, they require movement joints at twice the normal frequency.
Local overstressing of masonry	Local overstressing of masonry can be caused by narrow piers between windows and/or doors, around meter boxes and adjacent to corners. Simple design rules are included in *Approved Document A1/2* and cover the requirements for: • wall thickness • wall height • wall tie spacing • parapet heights • building proportions • building floor area • strength of bricks and blocks • buttress requirements • lateral support requirements. Where the rules cannot be followed, a Structural Engineer's design will be required. Note that *Approved Document A1/2* does not cover the design of lintels or beams.
Brick arches overstress adjacent brickwork	The stresses imposed by arch thrusts are outside the scope of *Approved Document A1/2* and should be checked by calculation.
Lightweight block inner leaf overstressed at beam bearings	A 150-200mm bearing is required for most lightweight lintels, therefore no openings with separate lintels can be closer than 300-400mm.
Lintels overstressed by loadings from concrete floor beams	All manufacturers of lightweight (steel) lintels emphasise the need for special consideration in situations where heavy point loads can occur. This is very common where concrete floor beams bear on lintels in external walls and, in these circumstances, a minimum of 3 brick courses is usually required over the lintels to spread the loads. This requirement should be checked in relation to floor-to-ceiling heights.
Inadequate restraint to walls from floors and roof	BS 8103: Part 1: 1986 *Structural design of low-rise buildings: Code of practice for stability, site investigation, foundations and ground floor slabs for housing* gives clear guidance on the restraint required to walls from the floors and the roof.

Design

COMMON DEFECTS	CRITERIA/GUIDANCE
Increased risk of water penetration	The risk of water penetration in periods of driving rain can be increased by using: • recessed mortar joints • low-porosity bricks • insulation-filled cavities • perforated bricks. The extent of the risk is also influenced by the site exposure. Refer to BSI Draft for Development 93: 1984 *Methods for assessing exposure to wind-driven rain* for a means of quantifying the exposure. Where the risk of water penetration is high, the quality of detailing and workmanship should be correspondingly high.
No lintels over small openings	There should be sufficient support to masonry over all openings, no matter now small. Galvanized steel angle may be used over openings of up to 450mm. The cut ends of the galvanized steel angle should be protected by bitumen paint.
Inadequate masonry over lintel to achieve composite action	To achieve sufficient compression, some lightweight lintels should be overlaid with a minimum amount of masonry. If in doubt, the lintel manufacturer should always be consulted.
Inadequate attention to mixed strength requirements for blocks	Special strength blocks may be required in certain locations at certain levels, for example at narrow piers at ground floor level. The correct identification of block types on site then becomes important. Consult the block manufacturer to check the identification system in use.

HAPM GUIDANCE

Designers should confirm that the building falls within the scope of *Approved Document A1/2* or identify specific shortfalls and substantiate the design for such instances. Lintels or beams should be designed or selected from appropriate proprietary components using manufacturers' design guidance where it is directly relevant. Scheme drawings should indicate clearly all restraint strap positions, block strengths and movement joint positions.

Details

COMMON DEFECTS	CRITERIA/GUIDANCE
Dpcs not fully considered	Dpcs should be carefully considered: • at steps • at staggers • at ground level changes • at horizontal cavity closures • around frames. Comprehensive details of things to be considered are included in BS 5628: Part 3: 1985 *Code of practice for use of masonry: Materials and components, design and workmanship*. Refer also to *Damp Proof Course Detailing* by Duell and Lawson and BS 8215: 1991 *Code of practice for the design and installation of damp-proof courses in masonry construction.*
Dpc/cavity trays lack stop ends, sealed laps and sealed corners	Note that the guidance on the installation of cavity fill advises that stop ends should be provided to the trays over lintels. At joints and corners, the dpc should be lapped by 100mm. All laps and corners should be sealed with the appropriate proprietary adhesive.
Cill details fail to shed water clear of building, and window/ door heads lack drips	It is important to recognise that if water shedding is not achieved, the durability of the bricks and mortar becomes much more critical. There are plenty of tried-and-tested cill details around. Drips should be provided unless the window and door heads are protected by the structure.
Seal between frame and wall not a formal 'designed' joint	Since a triangular mastic fillet is seldom applied properly, it is much better to design a recess to accommodate the mastic and to include a backing strip or bond break tape.
Restraint straps unable to work in compression and tension	BS 8103: Part 1: 1986 *Structural design of low-rise buildings: Code of practice for stability, site investigation, foundations and ground floor slabs for housing* gives a range of well-detailed, practical examples of the correct use of restraint straps.
Thermal insulation discontinuous at element junctions and around openings in walls	BRE *Thermal insulation: avoiding risks* gives guidance on generic solutions to discontinuous thermal insulation. However, this document does not consider practical problems, such as the fixing of frames where the cavity is closed by the insulation rather than by a brick or block return.
Movement joints too narrow or not full depth of leaf	Movement joints should be fully detailed and the nature of the compressible filler indicated. BS 5628: Part 3: 1985 *Use of masonry: Materials and components, design and workmanship* states that: • joints in clay brickwork should be a minimum of 16mm wide to accommodate expansive movement • joints in calcium silicate brickwork should be a maximum of 10mm wide to accommodate contraction of the brickwork. Some commonly used materials, such as bitumen-impregnated fibre board, can be too incompressible for use in expansion joints.

HAPM GUIDANCE

Sections through window head, cill and jamb, door head, jamb and threshold, ground floor/wall junction, intermediate floor/wall junction (all conditions) should be provided. In assessing the suitability of details, the exposure of the site should be taken into consideration.

Materials

COMMON DEFECTS	CRITERIA/GUIDANCE
Inadequate resistance of bricks to frost	Generally, bricks should be selected to suit the worst exposure condition unless more durable units are used for vulnerable details.
Corrosion of lintels	Evidence now available indicates that protection systems to some lightweight steel lintels can be significantly inadequate. Refer to BS 5977: Part 2: 1983 *Lintels: Specification for prefabricated lintels* which gives protection specifications and differentiates between lintels that are designed to function as an integral cavity tray and those that require an independent dpm. The fire performance of steel components should be considered. Many lightweight steel lintels require a cement/sand render on the back to achieve the necessary fire resistance.
Failure of timber frames	The BSI, BBA and PSA operate quality assurance schemes for windows and doors. Note that economy softwood windows may have a life as short as 10 years.
Premature rot in timber frames	Preservative treatment is important to promote the life of non-durable timbers in external joinery. Recent research has also identified poor end-grain sealing as a significant factor. Poor end-grain sealing can best be controlled by specifying that all external joinery should have copious amounts of waterproof adhesive on all faces of the joints. Few joinery manufacturers are prepared to use true WBP-quality adhesives in manufacture. Refer to BRE *Information Paper 20/87* for more guidance.
Plaster failure on high suction blockwork	Water retentive plasters should be specified for use on high suction backgrounds (HSB)
uPVC frame sections sag	There has been little feedback on problems with uPVC windows but the sagging of bottom sash members has been noted quite widely. Since this is probably due to the omission of reinforcement or reinforcement being a loose fit within the section, care should be taken when specifying reinforcement.
Dpcs too narrow	Dpcs should be wide enough to ensure they can extend fully through a wall. This avoids load concentrations at the wall face where the mortar bridges the dpc.
Inappropriate cavity fill material selected for prevailing exposure (fully filled cavities)	The various materials available for cavity fill have different exposure limitations. The BBA Agrément Certificate restrictions should be applied in every case.
Vibration compaction of low-density cavity fill	Blown glass fibre cavity fill can settle significantly if the wall is subject to impact, such as the removal of bricks by hammer and bolster or percussive drilling. Where settlement is likely or suspected, bricks should be removed at a high level to confirm the degree of settlement and to refill as necessary.
Corrosion of wall ties	The recommended zinc deposition of galvanized steel wall ties is now 940g/m^2, probably making the use of stainless steel more economic. Refer to BSI Draft for Development 140: Part 2: 1987 *Wall ties: Recommendations for design of wall ties* for performance requirements.
Mortar specification not enhanced for cills, copings, parapets and chimneys	Less permeable mortars, designation (i) or (ii), should be used. Refer to Table 15 in BS 5628: Part 3: 1985 *Code of practice for use of masonry: Materials and components, design and workmanship.*

Materials

COMMON DEFECTS	CRITERIA/GUIDANCE
Inadequate steel cover to precast concrete components	The amount of cover should be tightly specified and checks carried out on site using a reinforcement depth meter to ensure the specification has been adhered to.
Sulphate resisting cement not used where necessary	Sulphate resisting cement should be used: • below the level of the dpc where sulphates are present in the ground • where clay bricks with no limit on their salt content (FN and MN to BS 3921: 1985 *Specification for clay bricks*) are used in exposed conditions, eg: – parapets – chimney stacks – rendered walls – areas of severe exposure to driving rain.

HAPM GUIDANCE

Specifiers will not be required to adhere to specific durability criteria for components that can be replaced (eg frames) however the notional life of all components of the external wall will be recorded in the audit document, together with any maintenance necessary to achieve their lives. Inaccessible components, such as wall ties and lintels, should be expected to have a target life in excess of 30 years. Replacement of sealants around frames and claddings and to movement joints should be considered normal maintenance.

Workmanship

COMMON DEFECTS	CRITERIA/GUIDANCE
Inadequate care with reveal construction	BRE *Control of quality in design and construction of dwellings, Volume 3: An aid to site inspection* includes useful checklists for site checking of these and many other items. However, note that these lists have not yet been updated to take account of full cavity fill.
Dpcs not fully bedded	Full bedding above and below the membrane is essential.
Tray dpcs not bonded at laps	Tray dpcs generally require bonding in addition to lapping.
Cavities not cleaned	BS 8000: Part 3: 1989 *Workmanship on building sites: Code of practice for masonry* covers many of the problem areas but is often unspecific and theoretical. For instance using draw battens is recommended to ensure clean cavities, however this practice is so unwelcome on site as to be unrealistic. Generally, the preferred method of ensuring clean cavities is the temporary omission of selected bricks along the lines of trays allowing regular cleaning while the mortar is soft. Care should be taken that trays are not damaged by cleaning operations.
Wall ties too long or too short, omitted or misplaced	A shortage of wall ties is a common defect, especially the failure to increase the number of ties at reveals and movement joints. Wide cavities and short ties often combine to reduce the embedment. The embedment depths of wall ties should be a minimum of 50mm.
Bad winter working practices	Refer to BDA *Building Note 3* for guidance on bricklaying in winter conditions.
Frames not fully bedded	All window and door frames should be fully bedded in mortar along the length of the bottom member. Soft bedding is not recommended as it may allow a frame to sag and distort.
Heads of frames not fixed	Head fixing of frames is often omitted because it is not clear as to how it should be done and the fixing to proprietary steel or concrete lintels is difficult. Special clips or fixings may be required. Lintel manufacturers should be able to offer advice.
Movement joints not formed well	The application of mastic is best left to specialists. Common omissions include: • primers • bond break tape • backing strip • proper tooling.
Dpcs not extended to the face of a wall	It is better for the dpc to protrude and be cut back with a knife, than to be pointed over.
Confusion with blockwork strength variations	Care should be taken when selecting blockwork on site. To make selection easier, manufacturers are now coding different block strengths with colours or marks.
Lightweight blocks saturated	All masonry, but especially lightweight blockwork, should be covered when stacked.
Cavity widths vary due to poor setting out	Care should be taken at the setting out stage since where cavity widths vary, it is usually as a result of inaccurate setting out.
Lintels/beams not bedded	The thickness of the bedding should be taken into account when designing precast concrete components.

Workmanship

COMMON DEFECTS	CRITERIA/GUIDANCE
Cavity insulant badly cut and fitted	Cavity insulant should be cut, not torn, using a sharp knife not a trowel. Small pieces of insulant should be discarded and not used.
Cavity insulation left down temporarily forming a trough	Troughs formed by the cavity above the insulation will accumulate mortar and rubbish leading to an increased risk of water penetration. It is also difficult to push the insulation down into the trough. Insulation should always project from a cavity.

HAPM GUIDANCE

Builders and Supervisors should pay particular attention to these and similar common problems. The Technical Audit Unit Site Inspectors will undertake a 'quality count'.

EXTERNAL TIMBER FRAMED WALLS
Design

COMMON DEFECTS	CRITERIA/GUIDANCE
Inadequate allowance for vertical expansion of masonry and shrinkage of timber members	Brickwork can expand by 2.5mm per storey. Timber can shrink by 5mm per storey on drying out and under compression due to load. This means that in a two storey house a differential movement of up to 15mm should be accommodated at eaves and window openings.
Shortage of movement joints in brick veneer cladding	The brick veneer to a timber frame structure is less restrained than the outer leaf of a masonry cavity wall, therefore more vertical movement joints are likely to be needed. The rule-of-thumb in BRE *Defect Action Sheet 18* should be followed, ie a movement joint: • at every other separating wall for clay brickwork • at every separating wall for concrete blocks or calcium silicate bricks. A Structural Engineer's design will be required for any item outside the scope of *Approved Document A1/2.*
Piers too narrow	The piers may be overstressed or unstable if they are too narrow: • between windows and doors • adjacent to corners • between meter boxes. The resultant restricted areas of full height sheathing leads to less racking being available.
No lintels over small openings	There should be sufficient support over all openings, no matter how small. Galvanized steel angle may be used for openings of up to 450mm. The cut ends of the galvanized steel angle should be protected with bitumen paint.
Timber lintels overstressed by floor loadings or point loads	Timber wall frames should comply with the requirements of BS 5268: Part 6: 1989 *Structural use of timber: Code of practice for timber frame walls.* Timber components should comply with the requirements of BS 5268: Part 2: 1989 *Structural use of timber: Code of practice for permissible stress design, materials and workmanship,* and will generally be Grade SC3 or SC4.
Wall ties lack stiffness/ flexibility	Brick cladding should be connected to the timber framing with wall ties that have sufficient strength and stiffness to transfer wind forces to the timber framed wall. The wall ties should also have sufficient vertical flexibility to permit the vertical movement of the timber frame in relation to the brick cladding. Note that wall ties are normally spaced at 600mm horizontally to pick up a fixing on the timber studs.

Design

COMMON DEFECTS	CRITERIA/GUIDANCE
Insufficient resistance to external fire penetration	Walls on or within one metre of the (notional) boundary require fire resistance from both the inside and the outside. Tile hanging and other lightweight claddings will not normally achieve the necessary external fire resistance.

HAPM GUIDANCE

Timber framed construction is outside the scope of *Approved Document A1/2* and the adequacy of the design should be checked and certified by a Chartered Structural Engineer who is experienced in timber framed design. Guidance may be obtained from the TRADA publication *Timber Frame Construction* and from Section 6.1 in BS 5268: Part 6: 1988 *Structural use of timber: Code of practice for timber frame walls: Dwellings not exceeding three storeys.* Proprietary components should be justified by manufacturer's design literature. The site should be provided with assembly drawings showing frame connections, weathering and damp-proofing details, erection sequence and anchorage details of frame to substructure, quality and size of components, restraint strap positions, cavity barriers, tie requirements and preservative treatment.

Plasterboard is the most commonly used wall lining material but it cannot be relied upon solely to provide racking resistance. Refer to Section 6.1 in BS 5268: Part 6: 1988 for the limited contribution plasterboard may make to racking resistance in external walls and for bracing required and other limitations in separating walls.

Fire resistance of a timber dwelling should be considered in terms of the interaction of complete elements of construction and not to individual materials. Refer to BRE *Digest 214, 215, 218* and *230.*

Details

COMMON DEFECTS	CRITERIA/GUIDANCE
Dpcs not fully considered	Dpcs should be carefully considered: • at steps • at staggers • at ground level changes • at horizontal and vertical cavity barriers • around door and window frames. Comprehensive details of the criteria that should be considered are included in *Principle of Timber Frame Construction* by The Swedish Finnish Timber Council. Note that separating walls become partly external walls at steps and staggers.
Cill details fail to shed water clear of building, and window/door heads lack drips	It is important to recognise that if water shedding is not achieved, the durability of the bricks and mortar becomes much more critical. There are plenty of tried-and-tested cill details around. Drips should be provided unless the window and door heads are protected by the structure.
Seals around door and window frames unable to cope with differential movement	Since window and door frames are fixed rigidly to the timber frame, there should be provision for movement between them and any brick veneer. Joints should be designed for all frame perimeters. Refer to *Principle of Timber Frame Construction* by The Swedish Finnish Timber Council for guidance.
No clearance under verges and eaves to accommodate differential movement	Note that the differential movement is greatest at the top of the wall.
Cold bridging at floor perimeter	The risk of cold bridging with concrete ground floors is far greater in timber frame construction because the slab normally penetrates the cavity to the external building line when raft foundations are used. Refer to BRE *Thermal insulation: avoiding risks* which gives guidance for all common situations.
Services and obstructions in the cavity	Insulation should not be provided in any cavity behind the brick veneer. Services should be routed away from cavities and should not perforate any vapour check membranes.
Cavity barriers fail to seal the cavity	Cavity barriers are often inadequate because they fail to close the cavity fully. A degree of compressibility is important to accommodate any variation in the width of the cavity. Refer to BRE *Digest 214* and *215*.
No weep holes in cavities	Cavities should be drained sufficiently by the incorporation of weep holes below the level of the dpc.
Vapour control layers perforated by services	The responsibility for resolving the detailing of vapour control layers should not be delegated to site level.

HAPM GUIDANCE

Sections through eaves, window head, cill and jamb, door head, jamb and threshold, ground floor/wall junction, including holding-down details and intermediate floor/wall junctions, should be provided.

In assessing the suitability of details, the exposure of the site should be taken into consideration.

Materials

COMMON DEFECTS	CRITERIA/GUIDANCE
Insufficient frost resistance/ water penetration for bricks	All bricks and the pointing method used should be selected to suit the worst exposure condition, in conjunction with advice from the brick manufacturer.
Corrosion of wall ties	The recommended zinc deposition for galvanized steel wall ties is now 940g/m², probably making the use of stainless steel ties more economic. Refer to BSI Draft for Development 140: Part 2: 1987 *Wall ties: Recommendations for design of wall ties* for performance requirements.
Wall ties lack stiffness	Wall ties in timber frame construction should be stiff enough to resist the wind forces on the brickwork outer leaf but not so stiff as to hinder the differential movement between the frame and the brickwork. Specially designed timber frame ties should be used and fixed in accordance with the manufacturer's recommendations.
Corrosion of lintels	BS 5977: Part 2: 1983 *Lintels: Specification for prefabricated lintels* gives the requirements for protection systems to steel lintels.
Mortar specification not enhanced for cills, parapets, copings and chimneys	Less permeable mortars, designation (i) or (ii), can be used for lower absorption masonry units. Refer to Table 15 in BS 5628: Part 3: 1985 *Code of practice for use of masonry: Materials and components, design and workmanship* for guidance.
Joint failure of timber joinery components	Joint failure of timber joinery components can lead to: • premature rotting • inadequate weatherproofing • distortion of doors and windows. The BSI, BBA and PSA operate quality assurance schemes for windows and doors. Note that economy softwood windows may have a life as short as 10 years.
Inadequate cross-section/ fall to timber cills	In the absence of sub-cills, timber frame construction results in the need for wide timber cills which are liable to 'cup' if they do not have a sufficient cross-section. The section size is governed by the minimum thickness and minimum fall of the cill.
Limited life of weatherstripping/ glazing gaskets and movement joints seals, and inadequate allowance for differential movement in seals	Maintenance programmes should allow for the replacement of seals after a life of 5-8 years depending on the degree of exposure of the site. Movement joint seal materials and joint design should be carefully selected in conjunction with the manufacturer to ensure that the expected degree of movement falls within the capability of the system.

Materials

COMMON DEFECTS	CRITERIA/GUIDANCE
Inadequate preservation of timber frame components	Many errors can occur with timber preservation treatments. Refer to BS 5268: Part 5: 1989 *Structural use of timber: Code of practice for the preservative treatment of structural timber* and the *British Wood Preserving Association Manual* for recommendations. Mixed bundles of Redwood and Whitewood should be treated with the Whitewood treatment cycle. Preferably, all timbers should be treated to avoid the potential confusion if sole plates only are treated.

HAPM GUIDANCE

Refer to BS 5268: Part 5: 1989 and the *British Wood Preserving Association Manual* for recommendations for preservative treatments. A fungicidal and insecticidal treatment using organic solvents applied by the double-vacuum process is preferred. Specifications should identify treatment cycles for both Redwood and Whitewood timbers. The supplier of the timber components should confirm that the chemicals used in the preservative treatment are compatible with the connections and their coatings.

Workmanship

COMMON DEFECTS	CRITERIA/GUIDANCE
Nailing requirements for connections disregarded, and incorrect type of nails used	A site nailing schedule should be provided for every structural connection showing: • number and spacing of nails or staples • size and type of nails or staples • corrosion protection required • method of nailing. Refer to Table 2 in BS 5268: Part 6: 1989 *Structural use of timber: Code of practice for timber frame walls* for the nailing requirements to sheathing to give the required racking resistance.
Cavities not cleaned	It is best to leave out bricks to accommodate cleaning cavity trays and horizontal cavity barriers.
Bulging of wall panels due to wall ties omitted/misplaced	If the wall ties are bent on site, the bend should be as close as possible to the fixing. Nails should be stainless steel, phosphor bronze or silicon bronze and compatible with the wall tie material.
Cavity widths vary due to poor setting out	Traditional brickwork would normally be constructed with a tolerance of ±10mm in a 5 metre length and ±20mm in a 6 metre height. However, as timber can be cut to ±1mm in length and timber panels can be made to ±3mm on the horizontal and vertical dimensions, brickwork has to be constructed to much smaller than normal tolerances to ensure a good plumb wall built tight to cavity barriers.
Wall panels not adequately tied down	Holding down straps should not be slack.
Wall tie fixing nails miss studs	The wall tie fixing nails should be driven into the studs but often miss because the position of the studs is hidden by breather paper. To avoid this problem, marker tapes or lines can be used.
Wall tie frequency not increased around openings	The spacing of cavity ties should be increased around openings to one every 4 courses.
Movement gaps omitted under frames, under eaves and under gables	Bricklayers often fail to leave design gaps as they do not understand the problems of differential movement.

HAPM GUIDANCE

Builders and Supervisors should pay particular attention to these and similar problems peculiar to timber frame construction. The Technical Audit Unit Site Inspectors will undertake a 'quality count'.

COMMON DEFECTS	CRITERIA/GUIDANCE
Poor sound insulation	*Approved Document E1/2/3* and BRE *Digest 333* give the requirements and guidance for the provision for sound insulation. The minimum superficial weight of separating walls should be maintained at: • 375kg/m², including finishes, for solid brickwork • 415kg/m² for solid blockwork, or for brickwork or blockwork with a 50mm cavity • 250kg/m² for blockwork with a 75mm cavity • 120kg/m² for flanking walls.
'Mirrored' construction features create weaknesses in sound insulation	While a plasterboard lining to timber separating walls generally gives good sound performance, plasterboard on masonry frequently gives problems due to the absence of a wet plaster system to seal up any air paths through the masonry. Where possible, plan the shorter walls of opposing rooms to form the separating wall. The distance between adjoining openings on either side of the separating wall should be checked. If precast blockwork flues on separating walls will not have sufficient density, relocation of the flues should be considered. Services should be planned to avoid chasing a separating wall (especially back-to-back). Note that a wall separating a habitable room from a refuse chute requires a minimum superficial weight of at least 1320kg/m².
Condensation at cold bridges	Problems can arise at steps and staggers, and at the junction with any external flanking walls, due to the conflict between the requirements for high mass and low thermal conductivity.
Rain penetration	Problems can arise at steps, especially if the steps are small enough to make cavity tray installation difficult. Alternatively, cladding to small steps should be considered. A similar situation exists where the separating wall projects through the external wall or roof. Where possible, this design feature should be avoided.

HAPM GUIDANCE

Designers should confirm that the building(s) fall within the scope of *Approved Documents E1/2/3* and *B2/3/4*, or identify departures and substantiate with sound test results. Dry lined masonry separating walls will be highlighted as a risk by the Technical Audit Unit, particularly if they are of single-leaf construction.

Details

COMMON DEFECTS	CRITERIA/GUIDANCE
Airborne sound paths through wall	Air paths are common within the depth of suspended ground and intermediate floors, and in the roof space. Joists bearing in the wall from both sides should be staggered (on plan) to reduce 'unfillable' gaps. Where possible, joist hangers (of the restraint variety) should be used. Where this is not possible, the wall should be rendered within the depth of the floor. Areas in the roof space should be rendered and the wall behind the stair string should be plastered.
Bricks not laid 'frog up'	To avoid the walls having inadequate mass, bricks should be laid 'frog up'.
No headers in brick separating wall	A brick bond which includes headers should be specified as required by *Approved Document E1/2/3*.
Incorrect specification for wall ties	Cavity walls should be tied with wire butterfly ties and not strip ties which can increase sound transmission because of their stiffness.
Insufficient resistance to flanking sound	Flanking walls should generally have a minimum superficial weight of not less than 120kg/m².
Window/door openings too close	Permanent openings on either side of the separating wall should be at least 650mm apart (including windows with permanent ventilators), unless they are separated by a projecting masonry wall.
Condensation at cold bridges	When the separating wall is bonded to the outer leaf, a cold bridge is formed unless insulation is applied to the internal wall faces. Additional thermal insulation will be needed where dense blockwork is returned into the external wall.
Rain penetration	Care should be taken when detailing around a wall projected through the roof or the external wall. Where an adjoining pitched roof abuts a separating wall, the separating wall changes to an external wall and therefore requires cavity trays, unless it is clad or is small enough to be protected by an overhang.
Inadequate fire separation	The following fire separation measures are required by the *Building Regulations*: ● cavity barriers for continuous boxed eaves (BRE *Defect Action Sheet 7*) ● fire stopping to the top of the wall ● closing the cavity in the separating wall ● using roof coverings of a limited combustibility for a distance of 200mm either side of the separating wall (refer to Table A7 in *Approved Document B2/3/4*), unless the wall is taken up through the roof.
Confusion on site between blocks of different density	To avoid confusing different density blocks on site, the manufacturer's advice on the means of distinguishing the blocks should be sought.

HAPM GUIDANCE

Scheme drawings should clearly detail solutions at all 'key junctions' as defined in *Approved Document E1/2/3*. Arrangements at steps should show waterproofing provisions at the base and tops of the walls.

Materials

COMMON DEFECTS	CRITERIA/GUIDANCE
Blocks of inadequate density	In addition to complying with 'deemed-to-satisfy' masonry construction densities, test results are admissible for non-standard or proprietary constructions.
Incorrect wall ties in cavity walls	Butterfly wire ties which transmit less sound should be used rather than metal strip ties.
Fire stopping at head of wall not resilient	Rock fibre should be used, and not mortar, above and below the sarking and between the tiling battens. The rock fibre at the head of the wall should be thick enough so that compression prevents it being dislodged.

HAPM GUIDANCE

The audit will seek confirmation (for example by BBA Agrément Certificates) of densities of masonry components where relevant. Where the separating wall construction is not a 'deemed-to-satisfy' type, independent test results should be submitted.

Workmanship

COMMON DEFECTS	CRITERIA/GUIDANCE
Perpends not solid filled	It is crucial to ensure perpends are solidly filled with mortar. Incompletely filled perpends: • reduce the mass of the wall • leave airborne sound paths • leave fire paths. It is particularly important to fill the joints with mortar: • in the roof space • within the floor depth • around ends of joists, etc.
Use of incorrect density blocks	Where the separating wall becomes an external wall, particular care should be taken not to carry the block type for the external wall or the separating wall beyond the point of change (however, note that high density blocks may be required to be carried into inner leaf of flanking external wall).
Incorrect ties in cavity separating walls	Particular care should be taken to use only butterfly wire ties in separating walls, even if other types of wall tie are used elsewhere in the building.
Airborne sound paths at chases for services	Where the wall is to be chased, any accidental perforation of the wall should be made good to an adequate standard.

HAPM GUIDANCE

The Technical Audit Unit Site Inspectors will undertake a 'quality count'. Supervisors should be vigilant on the need for full mortar bedding and filling of perpends.

Design

COMMON DEFECTS	CRITERIA/GUIDANCE
Insufficient resistance to direct sound transmission	Separating wall panels should be structurally independent for each dwelling. Each separating wall panel (excluding framing) should have a minimum superficial weight of not less than $25kg/m^2$. Services should not penetrate the plasterboard lining. Surface-fixed electrical fittings and mini-trunking should be used on separating walls.
Insufficient resistance to flanking sound transmission	Only minimal lightweight cladding can be continuous across the ends of the separating wall.
Insufficient resistance to passage of smoke and flame	Cavity barriers and fire stopping should be provided at element junctions and perimeters.
Inadequate restraint	Care should be taken to achieve structural stability without preventing the two leaves of the separating wall from being acoustically independent.

HAPM GUIDANCE

Designers should substantiate the sound, fire and strength attribute of all designs in the absence of a 'deemed-to-satisfy' specification. There is a fundamental conflict of interests between sound insulation – which relies on structural isolation – and structural stability – which usually requires load sharing between dwellings (racking resistance of external walls and diagonal roof bracing are often shared facilities requiring good connection between the leaves of separating walls).

Details

COMMON DEFECTS	CRITERIA/GUIDANCE
Construction too light	For the construction of the timber framed separating walls to be of sufficient weight, the following should be used: • two layers of plasterboard, one each of 19mm and 12.7mm thick, or • three layers of 12.7mm thick plasterboard.
Air paths through wall	Services should not penetrate the separating wall. Socket outlet boxes should be surface-mounted. Plasterboard should be fixed with the first layer horizontal and the next layer vertical and having staggered joints. The wall within the roof space should be constructed in a similar way.
Fire paths through wall	Before the gable triangle is placed, it should be lined with plasterboard or proprietary fire resisting board to ensure continuity with the studwork in the roof space. Note that the gable triangle may already have been so lined by the manufacturer.
Cavity not closed at element junctions and perimeter	Cavity barriers should not form a rigid connection between the leaves. A flexible material should be used, such as wire reinforced rock fibre. Alternatively, where rigid material is to be used, it should be fixed on one side only.
Insufficient resistance to fire at top of wall in step and stagger configuration	Separating walls can become an external wall in step and stagger situations. In these circumstances, careful consideration should be given to the external fire resistance on the relevant boundary.
Inadequate flashing detail at step and stagger configuration	Conventional flashing details are not applicable in a step and stagger configuration because a brickwork facing is not practical at steps. Since the exposed part of the separating wall will normally be protected by a lightweight cladding, special details will be required.

HAPM GUIDANCE

Details will be required for all junctions of timber separating walls, ie to floor, to external wall, to roof, and at any step/staggers.

Materials

COMMON DEFECTS	CRITERIA/GUIDANCE
Mineral fibre quilt too thin, paper-faced or too light	Sound insulation quilt should be unfaced. Where the quilt is suspended within the cavity, it should be not less than 25mm thick, and not less than 50mm thick if it is fitted within the frames. The density of the quilt should be not less than $12kg/m^3$.
Ties between leaves too stiff to achieve isolation	Where using wall ties cannot be avoided, they should not be thicker than 14g.
Plasterboard nails too short	Plasterboard nails should be long enough to penetrate the frame members by about 30mm, therefore special long nails may be required when 2 or 3 layers of plasterboard are used.

HAPM GUIDANCE

Materials and components within a separating wall are inaccessible and generally not available for maintenance inspection. Therefore all components should have a maintenance-free life in excess of 35 years. All timber should be treated with preservative (against fungal and insect attack). All metalwork should be hot-dip galvanized.

Workmanship

COMMON DEFECTS	CRITERIA/GUIDANCE
Air paths through wall	All plasterboard joints should be tight-butted. The first layer of plasterboard should be 19mm thick and fixed horizontally with staggered joints. The next layer of plasterboard should be 12.5mm thick, fixed vertically and nailed through the first layer to the frame members. Where possible, the plasterboard should be applied to the gable triangle before putting it in place, since achieving the required completeness of cover against fire when the gable triangle is in place is very difficult, particularly at eaves level.
Inadequate fixing of plasterboard	All layers of plasterboard (whether second or third) should be fixed through to the timbers. Marking the lines of the studs makes fixing to the timbers easier.
Cavity barriers omitted or discontinuous	Cavity barriers are required: • at the intermediate floor level • at the upper ceiling level • at the top and ends of the wall • at the top of any openings. Care should be taken that the cavity barriers are tight-butted at joints, and in contact with the surrounding materials.

HAPM GUIDANCE

Very careful supervision is required and has to be timed to occur before the work is covered. Records of inspection are recommended as part of site quality assurance procedures. The National Federation of Building Trade Employees publishes books of checklists (for ticking) for timber frame construction.

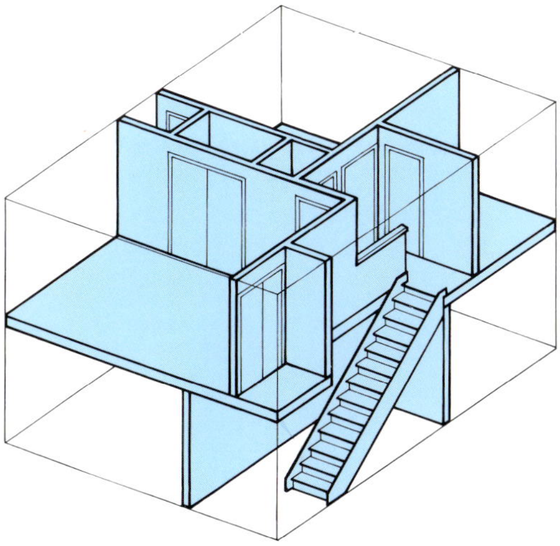

INTERMEDIATE FLOORS AND STAIRCASES
PARTITIONS

3

HAPM

HOUSING ASSOCIATION PROPERTY MUTUAL LIMITED

Design

COMMON DEFECTS	CRITERIA/GUIDANCE
Floors not designed for dead loads imposed by design	Partition loads and stair loads should be fully assessed. The dead load imposed by multiple layers of plasterboard for sound or fire insulation should be considered.
Timber floors 'springy' and liable to deflection and oscillation under live loads	*Approved Document A1/2, Part B*, gives some guidance on the sizes of timber floor members to be used in single family houses (species/grades/spans), but note that it gives only limited information on the necessary blocking and measures required to achieve full end restraint to joists. All ends of joists that are not built into a wall should be blocked or strutted beside the joist hangers, or in line with the wall plate. Blocking or strutting should be provided: • once, in spans of 2.5-4.5m • twice, in spans over 4.5m.
Excessive airborne or impact sound transmission through concrete separating floors	Excessive airborne sound transmission through concrete separating floors is a common problem where the timber deck is laid on beam and pot floors having no screed, since the resulting floor has many potential air paths. Grouting beam and pot floors is crucial to seal all air paths and to ensure that the whole floor acts as a single plate. Concrete beam or block floors with or without screeds may also lack the necessary mass for efficient airborne sound insulation. All sound tests and BBA Agrément Certificates should be checked.
Use of nominal 38mm timber for joists	Although *Approved Document A1/2* includes the use of 38mm wide timbers which, with permitted tolerances, are often even thinner, experience has shown that such thin members do not provide a nailing platform wide enough for either a timber floor deck or a plasterboard ceiling. Timber noggins nailed to the sides of the joists to increase their width may have to be used.
Incorrect thickness of timber board for joist spacing	*Approved Document A1/2* and BRE *Defect Action Sheet 31* recommend that tongued and grooved boards should be: • 16mm thick, up to 450mm joist centres, • 19mm thick, up to 600mm joist centres.
Screed too thin	Section 3 in BS 8203: 1987 *Code of practice for installation of sheet and tile flooring* recommends, as a minimum: • 40mm depth, for bonded screeds • 50mm depth, for unbonded screeds • 65mm depth, for floating screeds (light domestic use only) • 25mm cover of screed over any pipe, together with 250mm wide reinforcement over such pipes or conduits. However, it should be noted that floating screeds are not recommended due to the difficulty of compacting them properly.
Floor layouts redesigned on site to suit actual position of services	The position of all service pipes, ducts and drains should be carefully planned before the floor design is finalised.

Design

COMMON DEFECTS	CRITERIA/GUIDANCE
Blockwork partition walls built off timber floors	Blockwork walls built off timber floors tend to crack as the floor shrinks and deflects. Only timber and lightweight partitions should be built off timber floors.
Inadequate lighting of stairs	Lighting should be sufficient to highlight the top and bottom of all stair flights. Refer to BRE *Defect Action Sheet 53*.
Concrete stairs not designed as separating element	In some plans for flats, a precast concrete staircase separates one dwelling from another. Sound requirements should be met, especially the sealing of air paths between the stair flight and surrounding structure.

HAPM GUIDANCE

Sound test information should be provided for separating floor designs. Avoid heating pipes buried in screeds, unless in ducts. Stairs should conform to the requirements of BS 5395: Part 1: 1977 *Stairs, Ladders and Walkways: Code of practice for the design of straight stairs.*

Details

COMMON DEFECTS	CRITERIA/GUIDANCE
Location of blocking or strutting and noggins between joists not shown	All blocking or strutting should be shown on the drawings. The need for noggins at the following locations should also be considered: • plasterboard edges • light fittings • structural blocking • chipboard edges, etc.
Services runs not considered in relation to blocking or strutting and permitted hole or notch locations	Where heating or electric services are to be designed by others, the zones where pipes and cables can be routed should be shown. Refer to *Approved Documents A1/2 and B6*, and also to BRE *Defect Action Sheet 99*.
Inadequate detailing of support for partitions	Special attention is required for separating floors with resilient layers. Masonry partitions should not be built off floating screeds.
Inadequate detailing of chipboard flooring	Refer to BRE *Defect Action Sheet 31* and the *Chipboard Promotion Association Handbook* for guidance on whether noggins are required to support the edges of chipboard. Chipboard should not be allowed to cantilever and clearance at the perimeter should be allowed. Tongued and grooved joints should be glued.
Omission of thermal insulation over exposed soffits	Refer to BRE *Thermal insulation: avoiding risks* which gives guidance for all common situations.
Inadequate specification for and detailing of lateral restraint connections	Refer to the section on external walls and also to BRE *Defect Action Sheet 25* for guidance.
Poor lateral restraint at precast or prestressed concrete floors	Suitable details for lateral restraint are shown in BS 5628: Part 3: 1985 *Code of practice for use of masonry: Materials and components, design and workmanship* and BS 8103: Part 1: 1986 *Structural design of low-rise buildings: Code of practice for stability, site investigation, foundations and ground floor slabs for housing.*
Cold bridging at junction with external walls	Cold bridging at the junction with external walls is a particular problem with concrete floors and any associated upstand or downstand beams, and is always a problem with any cantilever projection. Local internal insulation should be provided. Refer to BRE *Thermal insulation: avoiding risks* which gives guidance for all common situations.
Insufficient bearing widths for concrete floor beams	BS 8110: Part 1: 1985 *Structural use of concrete: Code of practice for design and construction* requires a minimum bearing width of: • 100mm on Grade 30 concrete • 110mm on Grade 25 concrete • 105mm on masonry • 75mm on steel. Note that since a minimum bearing width of 75mm may well not be achieved on the flanges of steel beams (particularly the sloping surface of the lower flange), ledge angles may need to be welded to the beam.

Details

COMMON DEFECTS	CRITERIA/GUIDANCE
Heating pipes run in screeds	Pipes in screeds cause many maintenance problems. It is much better to provide pipework ducts with removable covers within the screed (as long as sound resistance is not compromised) or for pipework to run on walls.
No allowance for timber shrinkage where timber joists are notched into steel beams	Timber should be detailed to project 12mm above the steel section to accommodate future shrinkage of the timber.
Electrical services run above floor of flat below	Much damage occurs on site when electric cables for lighting to the flat below have to be passed through precast concrete floor systems. The resulting holes in the concrete floor may compromise the sound insulation. Maintenance can become a problem. It is much better to accommodate the electric cables in a batten-supported false ceiling.
Hand traps in handrails to stairs	Handrails should be designed with enough clearance from the adjacent structure to ensure a convenient handhold throughout the length of the rail. Ends of the rails should be detailed to avoid sleeve traps.
Handrails to stairs not strong enough	Refer to BS 6180: 1982 *Code of practice for protective barriers in and about buildings* for guidance.

HAPM GUIDANCE

A range of detail/assembly drawings are necessary for floors. Junctions with other vertical elements are normal requirements as well as assembly drawings for handrails and balustrades. Positions of blockings are best shown on joist layouts.

Materials

COMMON DEFECTS	CRITERIA/GUIDANCE
Joist hangers not matched to block strengths, or restraint-type hangers not specified where needed	Hangers should be marked with the manufacturer' s name and with the minimum compressive strength of brick or block for which they are suitable. Refer to the manufacturer' s literature and BS 5628: Part 1: 1985 *Code of practice for use of masonry: Structural use of unreinforced masonry.*
Confusion on timber strength requirements	Standardisation helps to avoid confusion but where this is not possible, drawings should indicate the locations required for the different stress grades of timber. Refer to BS 4978: 1988 *Specification for softwood grades for structural use* and BS 5268: Part 2: 1989 *Structural use of timber: Code of practice for permissible stress design, materials and workmanship* for guidance.
Allowance not made for difference between sawn and regularised sections when using span tables	Tables B.3 to B.28 in *Approved Document A1/2* give sawn timber sizes, except where noted otherwise. Therefore, allowance for this should be made when using regularised sizes (4mm undersized).
Timber or chipboard gets wet during construction, and chipboard gets wet in use	While the eventual use of chipboard may be in a dry environment, wetting during construction is almost inevitable. Therefore, as a minimum, chipboard should be Type C4 or C5 to BS 5669: Part 2: 1989 *Particleboard: Specification for wood chipboard.* To prevent chipboard getting wet while in use in kitchens and bathrooms, it should be covered with sheet vinyl with welded joints, and preferably with perimeter upstands.
Lateral restraint straps incorrect length or section, or countersunk head screws not specified for lateral restraint straps	Lateral restraint straps should be of minimum section 5mm x 30mm, and long enough to engage at least two joists. Since nails or round-headed screws may not 'fill' the predrilled holes in the straps, it is better to specify countersunk head screws.
Corrosion of metalwork	While a floor should be a protected environment, a minimum corrosion protection is required equivalent to BS 729: 1971 *Specification for hot-dip galvanized coatings on iron and steel articles*, that is a zinc coating of 460g/m² is required for steel articles under 5mm and not less than 2mm thick. Other specifications for the protection of joist hangers are included in BS 6178: Part 1: 1982.
Inadequate resistance to fire of metalwork connectors	The fire performance of metal connectors should be compatible with the fire resistance of the floor. Section 4.1 in BS 5268: Part 4: 1978 *Structural use of timber: Fire resistance of timber structures* states that steel 1mm thick, in conjunction with a 12mm plasterboard ceiling, should achieve 30 minutes fire resistance, and that 31mm of plasterboard would be required to achieve 1 hour fire resistance. BS 5268: Part 4: 1978 also suggests that 3mm steel would be required to achieve 30 minutes fire resistance without the plasterboard protection. However, BS 6178: Part 1: 1982 *Joist hangers: Specification for joist hangers for building into masonry walls of domestic dwellings* requires a minimum steel thickness of 2.5mm.
Trimmers different depth to floor joists	Normally, trimmers should be the same depth as the floor joists, with the width increased to suit the load.

Materials

COMMON DEFECTS	CRITERIA/GUIDANCE
Materials not of limited combustibility	Normally, stairs within protected shafts should be constructed in non-combustible materials. Refer to *Approved Document B2/3/4* for guidance.
Resilient layers too compressible	Long-term compression of the resilient layers provided for sound insulation may be a problem. Only materials specifically designed for sound insulation under screeds should be specified.

HAPM GUIDANCE

All materials should fall within the scope of *Approved Document B2/3/4*. Any chipboard used should be moisture-resistant and floor finishes should be fully detailed.

Workmanship

COMMON DEFECTS	CRITERIA/GUIDANCE
Holes or notches through joists not in permitted zone	*Approved Documents A1/2 and B6*, and BRE *Defect Action Sheet 99* give clear guidance.
Hangers not fully nailed to joists	Every hole should be nailed. Nails should be galvanized and a minimum of 38mm x 3.6mm square twisted.
Hangers not soundly bedded and tight to wall	Bearings for hangers should be level or bedding will be necessary. The back face of hangers should be hard against the support wall or timber.
Incorrect hangers for block strength	A joist hanger to BS 6178: Part 1: 1982 *Joist hangers: Specification for joist hangers for building into masonry walls of domestic dwellings* should have 2.8N/mm², 3.5N/mm² or 7.0N/mm² stamped on the side to denote the minimum crushing strength of masonry on which it rests.
Paired joists not equally loaded	Paired joists should be positioned so that each joist carries an equal share of the load. It is better to bolt the pairs of joists together to ensure good load sharing.
Omission of blocking and strutting	Drawings should indicate the positions of blocking and strutting. Blocking should include the blocks between the joists and between the joists and the walls.
Joists protrude into external wall cavity	Care should be taken to avoid joists protruding into the external wall cavity since they can provide a potential trap for mortar, and a possible path for moisture transfer.
Joists not solidly built into supporting walls	Design will inevitably require full end restraint. If in doubt, blocking should be added.
Lateral restraint connections executed inadequately	All restraint straps should be screwed (not nailed) to joists. Joists should be blocked at restraint points. Straps should not be bent up to bed joints. See Topic 2.2 and also refer to BRE *Defect Action Sheet 25 and 26*.
Chipboard flooring saturated	Particular care should be taken to avoid chipboard flooring getting saturated during construction since there can be a permanent loss of strength in the chipboard if it gets wet, even if it is Type C4 or C5. Tongued and grooved boards can shrink and cup.
Shrinkage of tongued and grooved boards	Tongued and grooved flooring should be tightly cramped prior to fixing to avoid shrinkage.
Chipboard not correctly jointed or fixed	Refer to BRE *Defect Action Sheet 31* and to the *Chipboard Promotion Association Handbook* for guidance.
Points loads not distributed (eg from bath feet)	Refer to BRE *Defect Action Sheet 32* for guidance.
Omission of insulation from exposed soffit floors	Since insulation is often fitted from below, care and patience is required to install it.
Screeds not sufficiently compacted or too dry	Refer to BRE *Defect Action Sheet 52* for guidance.

Workmanship

COMMON DEFECTS	CRITERIA/GUIDANCE
Gaps in sound deadening layer	Screeds can fill gaps and cause sound transfer across the isolating layer. Therefore, the isolating layer should be inspected before screeding, but note that it may also be liable to damage during the screeding work.
Joists not spaced as shown on drawings	The spacing of the joists shown on the drawings should be followed strictly and varied only if approved by a Structural Engineer.
Top tread or nosing of stairs left insecure or rocking	Joist hangers below the top tread of the stairs should be recessed into the trimmer to provide a flush surface for the tread.
Stair newel posts not fully secured	The newel post should be through-bolted to the floor joists.
Obstruction of stairs handrails by brackets or fixing screws, and handrails left unsanded	Continuous safe passage of the hand on the handrail should be achieved. Refer to BRE *Defect Action Sheet 54* for guidance.
Timber excessively bowed, twisted or cambered, and containing large edge knots or shakes	Defective timber should be rejected. Timber should be stress-graded before its arrival at site. The requirements of Clause 4.2 in BS 5268: Part 3: 1985 *Structural use of timber: Code of practice for trussed rafter roofs* form a useful guide to spring, bow, twist, wane, fissures, etc.
Inadequate nailing of multiple layers of plasterboard to joints	Multiple layers of plasterboard are very difficult to nail since the position of the joists and noggins will be obscured by the first layer of plasterboard. Therefore, it is very easy for the nails to miss the joist, especially with 38mm thick joists.
Fire separation not extended into floor or ceiling void	In certain situations where a fire division wall occurs, the floor and ceiling voids should be subdivided by means of cavity barriers. Refer to Appendix G in *Approved Document B2/3/4* for more details.

HAPM GUIDANCE

The Technical Audit Unit Site Inspectors will undertake a 'quality count'. Experience indicates that floor construction often suffers from 'lack of care' rather than 'lack of skill'.

Design

COMMON DEFECTS	CRITERIA/GUIDANCE
Upper storey partitions not aligned with partitions below, or supported inadequately	Masonry partitions should be supported on steel beams or taken down to their own foundation. The rule-of-thumb solution for lightweight partitions is to double the joists and spike them together, but calculations may be necessary to ensure the floor can support the partition load.
Layout locates door openings so that adjacent lintels achieve poor bearing	Two doors should not be located too close to a corner (on plan) where the partitions meet at right angles because this will leave insufficient wall to carry the lintel ends at the corner.
Non-loadbearing partitions can become 'loaded' due to deflection of floor above	Deflection of the floor above can result in an unacceptable load bearing on non-loadbearing partitions. It is best to make provision for movement at the heads of non-loadbearing partitions while still maintaining restraint, for example by incorporating a soft joint at the head of masonry partitions. Note that long span prestressed concrete floors are most vulnerable to permanent deflection due to long-term creep.
Inadequate protection of ground floor partitions by dpc	The dpc in ground floor partitions should be continuous with the slab dpm.
Insufficient sound insulation achieved by partitions	Lightweight partitions and door sets enclosing toilets are frequently unsatisfactory with regard to sound insulation. Sound insulation can be improved by using two layers of plasterboard and a heavyweight door.
Insufficient resistance to fire achieved by partitions to protected shaft	*Approved Document B2/3/4* gives the minimum periods of fire resistance for the elements of structure as follows: • Table 1.4 applies to dwelling houses • Table 2.1 applies to flats • Table 3.1 applies to institutional buildings.
Services located in thin masonry partitions	Chasing of a 50mm masonry partition to accommodate a pipe or cable does not leave much partition. Therefore, service runs should be carefully considered and defined, and not left to the whim of an operative. Cross-chasing should be avoided.

HAPM GUIDANCE

Where mandatory requirements apply, fire resistance, sound performance and strength should be substantiated.

Details

COMMON DEFECTS	CRITERIA/GUIDANCE
Lintels have inadequate masonry for composite action	Prestressed concrete lintels are usually designed to have 4 brick courses (or its blockwork equivalent) above.
Lintel bearings clash	Where adjacent doorways are close, there may be insufficient space to accommodate two lintel bearings. This is a particular problem at the corners of the room.
No lintels specified over small openings	Lintels (or galvanized steel angle) should be specified over minor openings to prevent cracking of blockwork and plaster.
Overloading of partitions	The layout plan should be checked to ensure that partitions are not overloaded, for example by carrying the ends of two lintels and the end of a trimmer.
Inadequate support to upper storey partition	Support for upper storey partitions should be specified. Two storey partitions should be blocked between the joists where these are parallel to the partitions.
Partitions built off floating screeds	Masonry partitions should not be built off floating screeds, however timber framed partitions may be.
No head restraint to partition	All partitions require head restraint and the method to be used should be clear on the drawings. Where deflection of the structure above the partition may impose a load on a non-loadbearing partition, the head restraint may need to be a sliding connection (soft joint).
Omission of cripple studs	The bearing requirements of the timber lintel design should be checked.
Structural noggin between studs not specified	Positioning of noggins should be clearly indicated.
Noggins for fixing socket outlets or kitchen wall units not specified	Fixing of electrical boxes with lateral fixings to vertical studs only is not good practice. Heavy items, such as kitchen wall units or WC cisterns, should be fixed to substantial noggins between vertical studs in predetermined locations.
Noggins at plasterboard joints not specified	To ensure that all free plasterboard edges are fully supported, noggins should be specified.

HAPM GUIDANCE

Details should be prepared for the head and base of all partitions to be interpreted in conjunction with partition layouts on which service positions and noggins should be indicated.

Materials

COMMON DEFECTS	CRITERIA/GUIDANCE
Incorrect specification for plaster	Refer to the British Gypsum *White Book*, BS 5492: 1977 *Code of practice for internal plastering* and BRE *Digest 213*. The need for water retentive plaster (HSB) should be checked.
Plaster suffers from poor adhesion, shrinkage or cracking	Blocks to be plastered should not be too dry otherwise moisture will be drawn from the plaster resulting in poor adhesion. If necessary, blocks should be wetted prior to plastering. Plaster mixes should not be too wet otherwise shrinkage cracking will be evident.
Cracking or loss of plaster adhesion at changes of substrate	Expanded metal lath or scrim should be used to bridge the junction between partitions. Alternatively, a movement joint should be detailed.
Incorrect plasterboard nails or incorrect nails for studwork and for fixing base rail	Refer to the British Gypsum *White Book* for guidance.

HAPM GUIDANCE

It is envisaged that most partitions, door linings and doors, if subject to normal use, will have a life in excess of the Major Defects Insurance Policy. Some specifiers of public sector housing require very 'robust' designs for partitions as a precaution against 'heavy booted' occupants. Perceived concern over noise nuisance within dwellings may generate a need for more heavyweight materials and components.

Workmanship

COMMON DEFECTS	CRITERIA/GUIDANCE
Lintels bearing on badly cut blocks or lintels supported on compressible packing	Blockwork should be arranged so that both ends of the lintels bear on full block and are levelled on a mortar bed.
Inadequate bonding at partition ends, especially where course mismatched or specified ties omitted	To prevent the reduction of mutual buttressing, care should be taken when bonding or tying the ends of partitions together.
Partition built off flooring spanning between joists	Partitions should not be built off flooring spanning between joists. Properly considered joist support should be provided. Partitions built off chipboard floors are particularly hazardous. For very lightweight partition systems, blocking between joists may be a satisfactory alternative, however, the adjacent joists may then need to be increased in size or doubled in number.
Partition built off floating screed	Ensure that partitions are not built off floating screeds in an unauthorised manner just to achieve greater screeding efficiency.
Plasterboard used after wetting or distortion due to poor site storage	To prevent the distortion of plasterboard by wetting, it should be stacked on a level surface off the ground either inside or outside (if protected sufficiently by tarpaulins). All damaged or distorted plasterboard should be rejected.
Studwork built up tight to deflecting floor above	To accommodate deflection of the floor above, a gap should always be left at the head of non-loadbearing partitions.
Inadequate fixing of plasterboard	Plasterboard should be nailed at no more than 150mm centres, vertically and horizontally.
Plasterboard not sealed prior to direct decoration	Plasterboard should be sealed with a proprietary wallboard sealer prior to decoration or wallpapering.
Floors overloaded by plasterboard stacks	To avoid overloading the floor, plasterboard should not be stacked in the middle of the floor but should be temporarily stored on edge around the perimeter of a room.

HAPM GUIDANCE

The Technical Audit Unit Site Inspectors will undertake a 'quality count' and they may note any unintended loading of partitions and any unintended support by partitions to other elements.

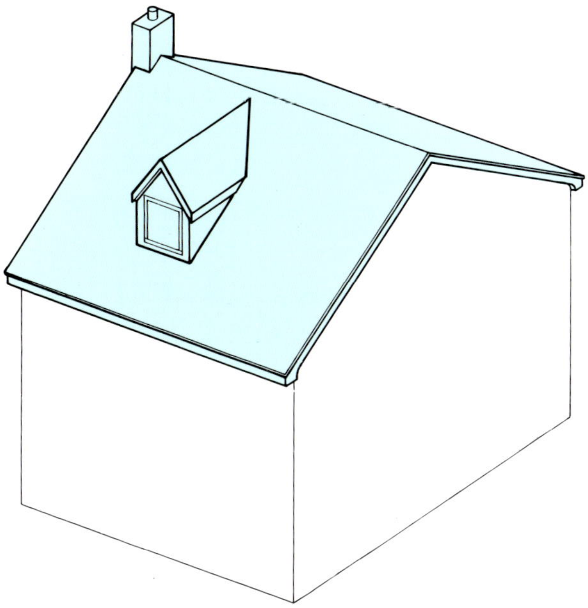

HAPM

HOUSING ASSOCIATION PROPERTY MUTUAL LIMITED

PITCHED ROOFS

FLAT ROOFS

DORMERS

CHIMNEYS AND FLUES

4

Design

COMMON DEFECTS	CRITERIA/GUIDANCE
Tiling not designed for prevailing exposure conditions	Wind loads (and suction) should be calculated as prescribed in CP 3: Chapter V: Part 2: 1972 *Code of basic data for the design of buildings: Loading: Wind loads.* The flatter the slope of a roof, the higher the suction. Increased suction should be allowed for: • at the periphery • along the ridge • around such features as chimneys. Designers should specify all nailing and clipping requirements precisely. BS 5534: Part 2: 1986 *Slating and tiling: Design charts for fixing roof slating and tiling against wind uplift* gives charts and tables for the tile fixing requirements for all conditions. Major manufacturers will provide free computer calculations for fixing requirements.
Different roof pitches abut at hip	Differences in adjacent pitches give differing heights of eaves for the same overhang. Varying the overhang to give a common fascia height can disrupt the common tiling gauge. An acceptable aesthetic solution can be achieved by setting out the tiles to suit the longest rafter length and accepting a reduced gauge on adjacent shorter rafters, but not the other way round. Any reduction in gauge below the minimum will not be accepted.
Gutters not designed for local storm conditions	The effects of local concentrations of discharge caused by funnelling (eg at valleys, around dormers, etc) should be checked. Refer to BS 6367: 1983 *Code of practice for drainage of roofs and paved areas* and BRE *Defect Action Sheet 55* for the design of roof drainage.
Roof configuration prohibits cross-ventilation of roof void	*Approved Document F* and Table 3 in BS 5268: Part 3: 1985 *Structural use of timber: Code of practice for trussed rafter roofs* give the minimum requirements for ventilator openings. See also BS 5250: 1989 *Code of practice for control of condensation in buildings.* High-level openings are required on monopitch roofs and recommended on steeply-pitched duopitch roofs. Refer to BRE *Defect Action Sheet 1* and *4.*
Trussed rafter roofs used for buildings of irregular shape	Trussed rafter roofs are most suited to buildings of simple plan form. It may be more straightforward to consider other systems of support for irregular shapes; possibly trussed purlins or traditional jack rafter and beam methods. However, trussed rafter design and manufacturing technology has improved in recent years and is now able to cope with a wider range of plan forms. The trussed rafter designer should be consulted at an early stage.
Compartmentation not extended into roof void	Dividing fire walls are generally required in the roof space to complete internal compartmentation, but conflicting ventilation requirements may have to be resolved.
No allowance for weight of water tanks	Tank loads can usually be included in truss design programmes. BS 5268: Part 3: 1985 requires trussed rafter designers to include for a water tank load of 900N (230 litre tank spread over 3 trusses or 300 litre tank spread over 4 trusses). Tank sizes in excess of this should be notified to the truss rafter designers.

Design

COMMON DEFECTS	CRITERIA/GUIDANCE
Bracing of roof trusses not fully defined	Drawings should be provided so that the Builder is not left in any doubt as to what is required. All necessary longitudinal binders and diagonal rafter bracing should be shown. Chevron bracing is required on the internal web members of trusses with a span greater than 8 metres and on monopitched trusses with a span greater than 5m. Refer to BRE *Defect Action Sheet 110* and BS 5268: Part 3: 1985.
Snow loads not accommodated	Some allowance for snow will probably have been included in a truss design programme. However, it is possible to get a build-up of snow or ice in times of continual freeze and thaw when gutters and outlets become blocked and for this to develop into solid masses of ice, especially in: • buildings of unusual shape • buildings which have roofs at different levels • buildings which have parapets. The medium-term allowable overstresses in the timber are then insufficient for this overloading and can result in the failure of the roof structure. The design of roofs where this is likely to occur should allow for such overloading. Refer to BS 6399: Part 3: 1988 *Loading for buildings: Code of practice for imposed roof loads*, and BRE *Digest 332*.
Whole roof removal by wind not anticipated by design	Restraint with a factor of safety of 1.4 should be provided against whole roof removal by wind uplift. Generally, there is sufficient dead weight in roofs over about 25° pitch, clad with tiles, to resist this uplift. However, in the following situations: • roofs of shallow pitch, clad with lightweight slates • roofs on high buildings • roofs in exposed areas additional restraint from the supporting structure should be provided by: • clipping the trusses to the wall plates, and • strapping the wall plates down to the structure.
Leaking valleys on shallow-pitched tile or slate roofs	It is very difficult to make a shallow valley sufficiently waterproof (remembering that the pitch of a valley is always less than the rafter pitch). Therefore, it is best to avoid valleys on roofs with a rafter pitch of below 22°.

HAPM GUIDANCE

Conventionally, timbered roofs of common or jack rafters supported by purlins may be designed by selection of timber sizes from the Tables in *Approved Document A1/2*. Trussed rafter and trussed purlin roofs are normally computer designed by trussed roof manufacturers. It is preferred that such manufacturers are members of the TRADA Quality Assurance Scheme for Trussed Rafters. The information to be given to the trussed roof manufacturer, and the information to be given by the trussed roof manufacturer, is set out in Section 10 in BS 5268: Part 3: 1985. Each party then understands its legal responsibilities. In all cases, the Technical Audit Unit will require a copy of the roof design and details.

Details

COMMON DEFECTS	CRITERIA/GUIDANCE
Ceiling insulation blocks eaves ventilators	Ventilation requirements should be clearly detailed and should include positive measures to prevent blocking by insulation. Refer to BRE *Defect Action Sheet 1* and *4*.
Inadequate thermal insulation and ventilation where ceiling is at rafter level	When the insulation has to be placed at rafter level, a clear air path at least 50mm wide should be maintained by placing an inert board between the insulation and the sarking. Refer to BRE *Thermal insulation: avoiding risks* which gives guidance for all common situations.
Rain penetration at eaves, verges, chimneys, abutments, etc	The following locations for flashings should all be fully detailed: • eaves • verges • chimneys • abutments with walls of adjoining houses • svps • rooflights. Refer to the *Lead Development Association Handbook.*
Water trapped in sarking, especially at eaves of low-pitched roofs	When the sarking felt is not supported behind the fascia board, it can sag allowing water to collect and leak into the building. It can be difficult to achieve sufficient support on low-pitched roofs since the supporting board for the sarking felt may cause a 'damming' effect. Refer to BRE *Defect Action Sheet 9* and BS 5534: Part 1: 1990 *Slating and tiling: Design*.
Valleys too narrow and poorly detailed	Valley gutters often block because they are too narrow for maintenance. The clear width of the valley should not be less than 125mm. Counter battens should be provided to give adequate support to the end of the tile battens.
Valley lining too flexible	The space between supports in the valley is generally more than the notional truss spacing. The valley lining board should therefore be thick enough to span without deflection and support construction or maintenance loads. However, if the valley lining board is too thick, it will cause tiles to 'kick-up', unless rafters are notched to accommodate (as shown in most manufacturers' brochures, but seldom done) or the lining is set level with the top of the rafters. When notching is proposed, the structural stability of the rafter should be checked.
Trussed rafters split by skew nailing to wall plate	To avoid skew nailing, truss clips should be specified to connect the trusses to the wall plates.
Roof tiling lifted by separating walls	The tops of separating walls should be finished 25mm below the underside of tiling battens. The remaining space should be filled with compressible non-combustible material to accommodate deflection of the trussed rafters. This acts as a fire stop and will also prevent hogging of the tiles along the separating wall line.

Details

COMMON DEFECTS	CRITERIA/GUIDANCE
Insufficient or excessive eaves overhang	Generally, an eaves overhang of 600mm from the bearing gives sufficient protection to the top of the wall and is within the capacity of the rafter. A small, flush eaves overhang will limit protection to the wall. Structural checks are required where the overhang is excessive (greater than 700mm) or where snow guards are envisaged.
Insufficient or excessive verge overhang	The verge tile oversail should be between 38-50mm. A projecting type verge with gable ladder and bargeboard gives the best wall protection but, as with the eaves, the structural implications should be considered.

HAPM GUIDANCE

Scheme drawings should show details of eaves, verges, steps, valleys, junctions with separating walls, flashings, abutments, roof bracing and tank supports.

Materials

COMMON DEFECTS	CRITERIA/GUIDANCE
Poor quality trussed rafters	Only metal gang nail plate trusses covered by BBA Agrément Certificates should be used. Truss manufacture can be very variable so it is best to require truss manufacturers to be members of the TRADA Quality Assurance Scheme for Trussed Rafters.
Inadequate preservative treatment to timber	All timber should be treated with an organic solvent fungicide and/or insecticide applied by a double-vacuum process. Such timber includes: • trusses • tiling battens • bracing • wall plates • fascias • soffit boards, etc. The treatment should leave a tinted finish so that untreated timber can be easily distinguished on site. The preservative treatment should be specified to suit the type of wood being used and the treatment cycles should also be specified (it is not sufficient to specify only the double-vacuum process). Note that most timber in roofs is European Redwood or Whitewood and that Redwood readily absorbs preservative solvents while Whitewood does not. Refer to the *British Wood Preserving Association Manual* and BS 5268: Part 5: 1989 *Structural use of timber: Code of practice for the preservative treatment of structural timber.*
Protective treatment to steel components has inadequate life	The following components, if mild steel, should be hot-dip galvanized after fabrication: • restraint straps • holding down straps • hip irons • verge clips, etc. Nailing plates, truss clips, etc should be fabricated from zinc coated, mild steel sheet to Class G275 which denotes a zinc deposition of 275g/m^2. Refer to BS 2989: 1982 *Specification for continuously hot-dip zinc coated and iron-zinc alloy coated steel: wide strip, sheet/plate and slit wide strip.* All nails (apart from tile fixings) should be galvanized. When wet, Copper/Chrome/Arsenic salts can set up corrosion of the galvanized plate fasteners, hence the Technical Audit Unit's preference for organic solvents. Note that some insulation materials can also cause corrosion of plate fasteners.
Tiles slip due to corrosion of nails or clips	Silicon bronze or stainless steel nails are preferred for their durability, however aluminium alloy nails to BS 1202: Part 3: 1974 *Specification for nails: Aluminium nails* have been used successfully in normal exposure conditions in this country since 1950. Tile clips and eaves clips should be compatible with the nails and may be stainless steel, aluminium alloy or plastics (acetal co-polymer).

Materials

COMMON DEFECTS	CRITERIA/GUIDANCE
Fibre-based sarking felt degrades in ultra-violet light	The standard Type 1F sarking felt to BS 747: 1977 *Specification for roofing felts* is prone to disintegrate where dressed into gutters. BRE *Defect Action Sheet 9* recommends the provision of a supplementary strip of dpc-quality material at eaves and verges.
Flashings become brittle and split	Cheap materials tend to be less durable. Proprietary economy materials will have a nominal life recorded by the Technical Audit Unit. Lead flashings should be Code 4 or 5.
Inadequate durability of tank platforms	To avoid problems with condensation from the tank surface, chipboard or ordinary quality plywood should not be used. Refer to BRE *Defect Action Sheet 43.*
Mesh for ventilators too small or too large	The mesh for ventilators should be small enough to prevent the entry of birds and insects but large enough not to become blocked by debris, corrosion or paint. Stainless steel, expanded metal lathing to BS 1369: Part 1: 1987 *Steel lathing for internal plastering and external rendering: Specification for expanded metal and ribbed lathing* gives a mesh size of about 3-4mm recommended by BS 5250: 1989 *Code of practice for control of condensation in buildings.*
Inadequate drainage of rooflights	Drainage for condensate from rooflights should be provided.
Uncertain performance of dry verge or ridge systems	Due to the limited experience of the durability of plastics verge and ridge components, their replacement should be anticipated during the life of the building.

HAPM GUIDANCE

All materials should conform to the requirements of the relevant British Standards or should be supported by a BBA Agrément Certificate. Most roof coverings will have a life extending beyond the expiry date of the Major Defects Insurance Policy. Some roof coverings, such as artificial slates and concrete slates, may be qualified. It is recommended that truss manufacturers are members of the TRADA Quality Assurance Scheme for Trussed Rafters.

Workmanship

COMMON DEFECTS	CRITERIA/GUIDANCE
Half-tiles used at valleys and at verges	One-and-one-half tiles or slates should be used at valleys and verges, not half-tiles or slates which have no body in the remaining portion once they are cut to shape. The requirements for nailing and clipping should be followed strictly.
Inadequate fixing of ridge and hip tiles	Gale damage has shown how easily ridge and hip tiles can become detached. Mechanical fixing of ridge and hip tiles is preferred. Some manufacturers of traditional ridge tiles now cast-in a wire that may be fixed to the ridge board in exposed locations.
Use of incorrect mix for tile bedding mortar	The only suitable mix for mortar for tile bedding and pointing recommended in BS 5534: Part 1: 1990 *Slating and tiling: Design* is 1 : 3, cement : sharp sand.
No support to ends of tile battens	The ends of tile battens should be sawn square and butt-jointed centrally on the rafter so that sufficient bearing and nailing can be provided to each end of each batten. This requires particular care when the rafters are less than 38mm wide.
Multiple battens jointed over one rafter	Joints should be arranged so that not more than one batten in four is jointed on any one rafter. Additional counter battens should be provided at valleys to support the batten ends.
Sarking too taut or too limp	Sarking felt should not be stretched too tightly over rafters because of the danger of tears developing when it is nailed, and water can be trapped by the batten. Where the sarking felt is too limp, there will be the danger of water not draining away properly and ponding.
Sarking not watertight	Sarking should be cut neatly and dressed around features, such as vent pipes and chimneys, and should be dressed down into gutters. Refer to BRE *Defect Action Sheet 10*.
Flashings not fixed	Flashings should be chased and wedged a minimum of 25mm into brick walls at steps or staggers and chimneys. Wide flashings will require additional fixings to provide restraint against wind uplift.
Trussed rafters not stored properly or mishandled	Trussed rafters should be stored clear of the ground, in the vertical plane and with suitable level bearing packs at the bearing positions. Roofing should be completed within one month of truss fabrication. Refer to BRE *Defect Action Sheet 5*. Trussed rafters should be lifted only in the vertical plane. Modifications to trussed rafters at site are not permitted.

Workmanship

COMMON DEFECTS	CRITERIA/GUIDANCE
Trussed rafters erected out-of-plumb and with members distorted	Appendix C in BS 5268: Part 3: 1985 *Structural use of timber: Code of practice for trussed rafter roofs* suggests a method of erection for trussed rafters. Trussed rafters should be: • maintained in the vertical position • parallel to each other • at the correct spacing • restrained against lateral movement. The maximum deviation from the vertical is specified in Table 11 in BS 5268: Part 3: 1985 (for instance for a 2 metre high truss, the maximum deviation from the vertical should be 15mm). When trussed rafters are fabricated, the members should be free from distortion within specified limits. After erection, a maximum bow of 10mm may be permitted in any member, provided that it is secured firmly enough to prevent the bow from increasing.
Inadequate bracing of trussed rafters	Diagonal rafter bracing should normally extend at 45° from as close to the ridge as is practicable at one end of a roof to the wall plate. Longitudinal binders should then be fixed to all trusses at ceiling and rafter level at node points. Each member crossed by a bracing should be nailed twice using 3.35mm diameter x 65mm long galvanized round wire nails. Refer to BRE *Defect Action Sheet 83* and *84*.
Binders do not abut walls, and no noggins provided at restraint straps	The object of constructing a fully-braced roof structure is to provide lateral restraint to the external walls. This lateral restraint can be achieved only if all the binders and restraint straps are packed tightly against the walls, and noggins are fixed between trusses at the strap locations.
Binders, bracing or truss members cut out	Cases have been found where vital structural members have been cut away to clear svps or flues, or to construct loft extensions. In no circumstances should any modifications or repairs be made to any trussed rafter without the specific approval of a Structural Engineer.
Inadequate splicing of binders or bracing	Where bracing members have to be provided in more than one piece, say due to the difficulty of access, they should be jointed over at least two trussed rafters.
No preservative treatment to cut ends of timber members	All ends of timber members cut at site should receive a preservative treatment applied by brush and equivalent to the specified treatment to suit the species of wood being used.
Damage to dry verge or ridge components	Since damage may impair the weather resistance of these dry systems, care should be taken when handling the components.
Bearing lengths for ends of trussed purlins too short	To avoid compromising the lengths of the bearings when trussed purlins are to be supported by shoes or by a corbel, care should be taken to ensure the setting out and construction of the support walls is accurate.

HAPM GUIDANCE

Trussed rafters are highly engineered components and satisfactory performance can be ensured only by proper attention to the specification and selection of material, adequate control of production, satisfactory storage and handling and correct erection procedures and fixing. The Technical Audit Unit Site Inspectors will review erection standards and may check roofs for tolerances.

Design

COMMON DEFECTS	CRITERIA/GUIDANCE
Ponding of water	BS 6229: 1982 *Code of practice for flat roofs with continuously supported coverings* specifies that the fall of a finished roof should be: • 1 to 60 for aluminium, copper, zinc and lead coverings • not less than 1 to 80 for most other types of covering. However, ponding of water often results from neglecting the deflection of the roof members themselves induced by the self-weight of the structure. Therefore, in practice, a design fall of 1 to 40 should be used to accommodate this deflection.
Condensation in cold deck type roofs	The risk of condensation in a cold deck roof is high and difficult to predict. Even when the roof is correctly designed, condensation may still occur due to the lack of cross-ventilation in windless conditions. Where a free ventilation area of 0.4% of the roof area cannot be achieved, the design should be changed to give a warm or inverted type roof. Refer to BRE *Defect Action Sheet 59.*
Condensation within the construction	An efficient vapour control layer with sealed laps should be specified immediately below the insulation. The vapour control layer should be fully supported with all the joints bonded. The vapour control layer should be turned up at the edge of the insulation and bonded to the weatherproof membrane.
Water trapped in the construction	In the case of in-situ concrete roof construction and cementitious screeds, sufficient time should be allowed for the drying out of the water used in construction. If sufficient time is not allowed for, large quantities of moisture will be trapped below the vapour control layers or weatherproofing and this can often be mistakenly interpreted as condensation or water penetration. A method of allowing drying out after the roof covering is laid should be provided.
Loss of insulation value in inverted type roofs due to rain cooling	The designed thickness of insulation should be increased by 20% to allow for the intermittent heat loss in winter when rainwater flows over the relatively warm weatherproof insulated membrane.
Whole roof removal by wind not anticipated by design	Restraint with a factor of safety of 1.4 should be provided against whole roof removal by wind uplift. In the case of lightweight roofs, or roofs on high buildings or in exposed areas, restraint should be provided by: • strapping the wall plates • providing ballast of either 20mm aggregate or paving slabs.
Sound insulation not considered	A lightweight roof design will have limited sound reduction properties. All local sources of sound nuisance should be considered and accommodated, such as low flying aircraft or nearby road or rail traffic.
Inadequate roof drainage	Drainage should be in accordance with the requirements of BS 6367: 1983 *Code of practice for drainage of roofs and paved areas.* A design rate of rain fall of 75mm/hour is generally satisfactory. Drainage should be designed so that in the event of a blocked outlet, overflow will occur safely outside of the building rather than into the building.

Design

COMMON DEFECTS	CRITERIA/GUIDANCE
Vapour control layers laid directly on corrugated metal deck	It is essential that the vapour control layer below the insulation in a warm deck roof is laid 'fully bonded' (hot pour and roll) on a deck which provides uninterrupted support. This means that the vapour control layers should not be applied directly to a corrugated metal deck.

HAPM GUIDANCE

A flat roof is a composite unit and designers should recognise that the long-term performance of each material is interdependent upon the other materials involved. Refer to BS 6229: 1982 *Code of practice for flat roofs with continuously supported coverings.* Refer also to BS 6399: Part 3: 1988 *Loading for buildings: Code of practice for imposed roof loads* for loadings on flat roofs. There are many good guides to flat roofing. The Technical Audit Unit commends the *Technical Guide to Flat Roofing* published by the PSA. Reference should also be made to BS 5250: 1989 *Code of practice for control of condensation in buildings.* The cold deck type of roof has been the most commonly used type of construction for dwellings and is the most troublesome. Its use should be avoided wherever possible.

Details

COMMON DEFECTS	CRITERIA/GUIDANCE
Cross-ventilation route restricted	The clear depth of any ventilation void in a cold deck roof should not be less than 50mm. Openings for ventilation should be provided at each end of the cavity. Where the width of the opening is more than 10mm, it should be protected by a mesh of size 3-4mm to prevent the entry of insects and small birds.
Water leakage around vertical perforations	Soil vent pipes, roof ventilators, handrail standards, etc should be efficiently weathered with the waterproofing membrane taken up around the projection. The projection should be provided with a cover flashing or hood.
Perimeters of felt roof do not shed water clear of fascia	Welted drips of mineral-surfaced roofing material form a traditional edge detail and can be formed on timber battens to throw the rainwater clear of the fascia.
Water penetration at junction of flat roof and wall	Frequently, there is differential movement at the junction of flat roofs and walls which can lead to water penetration. The flat roof covering should have an integral upstand of about 150mm fixed only to the deck. The junction should be covered with a lead or felt flashing from the higher level. Refer to BRE *Defect Action Sheet 34*.
No provision for movement in long roofs	Where long lengths of terraced dwellings have a flat roof, there should be joints at approximately every other party wall to accommodate thermal, moisture and structural movements. The twin kerb upstand type of joint is preferred, with a capping of aluminium, copper or galvanized steel.
Differential movement between roof covering and deck	Differential movement between the roof covering and the deck can cause failure of the roof covering due to daily reversals of stress. For mastic asphalt roof coverings, independent movement of the asphalt is accommodated by the provision of a loose-laid isolating membrane under it. Recessed gutters formed in the asphalt are best avoided as they tend to 'lock' the asphalt which can lead to subsequent cracking. Built-up bituminous felt coverings are normally bonded (intermittently or continuously) or mechanically fixed to the deck. Bending, tension or compression and shear stresses are therefore developed in the covering and its connections due to the restraint imposed. The roof covering should be capable of resisting such frequent movements and stresses for its design life. Sheet metal roof coverings can normally accommodate differential movements because of the deformability of closely-spaced seam joints and clips. Refer to BRE *Defect Action Sheet 33*.
Inadequate fire stopping at junction of roof and separating walls	*Approved Document B2/3/4* gives the acceptable forms of construction for fire stopping.
Condensation on rainwater pipes in roof void	If uninsulated downpipes pass through an inadequately ventilated roof void, condensation may form on the pipes when cooled by rain. This defect can be avoided by providing thermal insulation to the rainwater pipes and by improving the ventilation to the roof void.

Details

COMMON DEFECTS	CRITERIA/GUIDANCE
Water penetration at coping	Where possible, all materials and components should conform to a British Standard. For materials and components not so covered, evidence of their satisfactory performance and suitability for use in particular situations should be obtained and carefully assessed and should be backed by a BBA Agrément Certificate. It is difficult to waterproof a parapet or flat roof junction where the coping to the parapet is close or level with the upstand height of the flat roof since such an arrangement prohibits the proper use of stepped cavity trays.
Wind scour of gravel ballast	Loose gravel may be provided to protect a bitumen felt membrane from ultra-violet degradation or to anchor the roof against wind uplift. The wind speed required to shift the gravel depends on: • the size of the stones • the aerodynamic flow over the stones, which in turn depends on: – location – exposure – shape of the building – proximity to other buildings. Clause 10 in BS 6229: 1982 *Code of practice for flat roofs with continuously supported coverings* recommends a minimum gravel size of 20mm. Refer also to BRE *Digest 311*.
Soil vent pipes, flues and rainwater pipes located too close to parapets, other upstands and other pipes	Where a vertical pipe perforation occurs too close to a vertical element, it will be very difficult to work any asphalt or dress any felt between the two. A working space distance of 300mm should therefore be allowed.
Failure of felt roof coverings at angles	In most roof surfacing materials, changes of direction of 90° should be avoided; the introduction of fillets will reduce the changes to 45°. Refer to BRE *Defect Action Sheet 34*.
Localised surface degradation of asphalt near flues	Where chimney outlets are sited near to roof level, flue gases can cause degradation of the asphalt. Raising the chimney, say one metre higher, would result in the exhaust being diluted over a wider area.
Condensation at cold bridges	The overlaps for roof and wall insulation should be specified. The paths of greatest heat loss should preferably provide a U value not greater than 1.2W/m^{2}degC.
Roof plan of asphalt covered roof includes internal angles	Stress cracking is common in asphalt at internal angles (on plan). Where it is not possible to avoid using internal angles, the addition of a movement joint should be considered.
Rainwater outlet left above level of roof	Where some deflection in the deck is inevitable, a flexible or sliding connection should be included in the rainwater downpipe.

HAPM GUIDANCE

Scheme information should include details of all roof perimeter conditions, all abutments, rainwater outlets, pipe and flue perforations and kerbs to rooflights. In cold roof designs, the position and extent of ventilation openings should be indicated.

Materials

COMMON DEFECTS	CRITERIA/GUIDANCE
Rot in structural timbers	As a precaution, all timber should be pretreated against fungal and insect attack.
Chipboard and strawboard decks disintegrate if allowed to become wet	Exterior-quality plywood or wood wool should be used.
Corrosion of metal hangers or straps connectors on fixings	All mild steel in roofs should be protected with hot-dip galvanizing or its equivalent.
Vapour control layers or checks too flimsy or too permeable	Vapour control layers should be resistant to perforation and to the passage of water vapour.
Inadequate performance of rooflights	Inadequate performance of the rooflights can lead to: • inadequate weathertightness • condensation • lack of security • external fire penetration requirements not being met. Rooflights are not covered by a British Standard but, where possible, rooflights with a BBA Agrément Certificate should be used. Appendix B in *Approved Document B2/3/4* contains a concession for the fire rating of rooflights made from plastics material.
Inadequate durability of flashings	Flashings should be formed from metal such as lead (Code 4) or copper (0.5mm thick). Potential problems arising from using dissimilar metals should be considered.
Fatigue failure of weatherproofing membrane	When insulation is placed directly beneath a weatherproofing membrane, surface temperature fluctuations in the membrane will be increased. The thermal stresses are greater when using higher levels of insulation. For built-up membranes, a high tensile membrane should be specified, such as Type 5 to BS 747: 1977 *Specification for roofing felts.* All felts and asphalt require a suitable solar reflective finish, such as: • a light-coloured mineral aggregate • solar reflective paint • metal foil • concrete tiles or slabs.
Degradation of insulants in inverted type roofs	Degradation of the insulating layer can be caused by: • freeze and/or thaw cycles • wetting • exposure to ultra-violet light. To avoid such degradation, a frost-resistant insulant with low water absorption should be specified. The ballast needed to resist wind uplift should provide enough protection against ultra-violet light.

Materials

COMMON DEFECTS	CRITERIA/GUIDANCE
Incorrect specification for grade of asphalt	For roofs which have limited access, roofing-type asphalt should be used, that is: • for limestone aggregate, Type R988 to BS 6925: 1988 *Specification for mastic asphalt for building and civil engineering (limestone aggregate)* • for natural rock aggregate, Type R1162 to BS 6577: 1985 *Specification for mastic asphalt for building (natural rock asphalt aggregate).* For roofs which incorporate paved areas, one of the following should be used over a layer of Type R988 asphalt: • asphalt to BS 1446: 1973 *Specification for mastic asphalt (natural rock asphalt fine aggregate) for roads and footways* • asphalt to BS 1447: 1988 *Specification for mastic asphalt (limestone fine aggregate) for roads, footways and pavings in buildings.*
Thermal movement of lead flats restrained by 'glueing' effect of felt underlay	Not all sheathing felts are suitable as underlays to lead flats. For example some bitumen underlays can result in the lead flats becoming attached to the deck and then cracking due to movements of the deck. The specification for the sheathing felt should be checked with the Lead Development Association.

HAPM GUIDANCE

All materials and components should, if possible, conform to a British Standard. For materials and components not so covered, evidence of their satisfactory performance and suitability for use in particular conditions should be obtained and carefully assessed and should be backed by a BBA Agrément Certificate.

Workmanship

COMMON DEFECTS	CRITERIA/GUIDANCE
Overloading roof by stacking construction materials and equipment	To avoid overloading the roof, only material required immediately should be placed on the roof. Rolls of roofing material should not be stacked or stood close together. Bitumen boilers should not be placed on the roof unless the loading has been checked by a Structural Engineer.
Insufficient drying time allowed for water used in construction of concrete roof slabs and screeds	When sufficient time is not allowed for the water used in construction to dry out and the waterproof membranes are applied too soon, the building may suffer from problems of damp for a considerable time. Ideally, one month should be allowed for the drying out of each 25mm thickness of concrete slab and screed. If this time is to be reduced, the Builder should be asked to produce moisture content readings for the slab and screed before being allowed to proceed.
Perforations in vapour control layers	All edges in the vapour control layers should be sealed. Holes formed for vent pipes, flues, handrail standards, etc should be sealed carefully. Any tears or accidental damage should be made good to achieve a high vapour resistance.
Moisture-sensitive decking saturated during construction	Particular care should be taken to prevent moisture getting trapped in a wood wool or timber deck during construction otherwise the deck elements can swell. When this trapped moisture dries out, the decking sheets shrink at the end joints. Over time, this cyclical thermal expansion and contraction will cause fatigue and splits in the felt.
Lead flashings used in too long lengths	Lead flashing should not be laid in too long a length because the lead pulls and splits under thermal movement. The recommended limit is 1.5m. Refer to *Lead Sheet in Building* published by the Lead Development Association for guidance.
Omission of welt from bottom edge of cover flashings	Moisture penetration can occur via the capillary path between the back of the flashing and the face of an asphalt upstand. A turned-in welt should be used to break this capillary path.
Overheating of asphalt	No grade of asphalt should be heated to a higher temperature than 230°C. Thermometers should be used to check the temperature.
Felt laid in cold, wet or windy weather without protection	In cold weather, rolls of bitumenized felt should be warmed before they are unrolled or they will tend to crack (hence the lower limit of 5°C in most specifications for work in cold weather). The general problems of laying bitumenized membranes in cold weather can be overcome if heat and shelter are provided to the work area. Moisture prevents the adhesion of bitumen to any surface. Refer to BRE *Defect Action Sheet 63*.

FLAT ROOFS

Workmanship

COMMON DEFECTS	CRITERIA/GUIDANCE
Omission of sand rubbing of asphalt	Sharp sand should be rubbed over all the asphalt surfaces immediately after it is laid and while it is still warm.

HAPM GUIDANCE

Roof coverings and their substrates should be laid in accordance with the manufacturer's instructions and with the relevant code of practice for the particular material. Refer to CP 144: Part 3: 1970 *Roof Coverings: Built-up bitumen felt* and CP 144: Part 4: 1970 *Roof Coverings: Mastic asphalt.* Builders and Supervisors should pay particular attention to the above common defects and to construction quality at all details. The Technical Audit Unit Site Inspectors will undertake a 'quality count'.

Design

COMMON DEFECTS	CRITERIA/GUIDANCE
Inadequate consideration of condensation	The geometry of dormers can be complex and problems arise in achieving continuity of insulation and vapour control layers. The dormer should not impede the ventilation of the main roof. Ventilation on the cold side of the insulation has to be maintained to the outside air. *Approved Document F2* and Table 3 of BS 5268: Part 3: 1985 *Structural use of timber: Code of practice for trussed rafter roofs* give the minimum requirements for ventilator openings. Pitched dormers may require ridge level ventilators. Refer to BRE *Thermal insulation: avoiding risks* which gives guidance for all common situations.
No rainwater collection from roof of dormer	Some guttering to the dormer should be provided rather than subjecting the vertical elements of the dormer to high concentrations of run-off.
Tiles or slates too big	Since a dormer roof is normally smaller than the main roof, large tiles or slates should be avoided.
Concentrations of rainwater discharge	Dormers change the normal roof rainwater discharge and cause concentrations of run-off which can make normal gutter design inapplicable. A design rate of rainfall of 75mm/hour is generally satisfactory. Refer to BS 6367: 1983 *Code of practice for drainage of roofs and paved areas* for the design of roof drainage and BRE *Defect Action Sheet 55*.
Dormers too frail	Dormers will often be located in negative pressure zones created by the flow of wind over the main roof. Therefore, the coverings to flat roofs should be carefully fixed down and the framing should be restrained sufficiently against wind uplift. The span of timber framed dormers in trussed rafter roofs is usually too small to justify the expense of fabricated frames and the dormer will need to be designed as a traditional roof with timber posts and purlins tied down to the main structure. The dormer frame should be completely independent of the window frame and an adequate lintel should be provided over the window.
Design does not allow for accessibility for maintenance	Dormers normally include items such as valley gutters or flat roofs as well as painted softwood, which all require regular maintenance. Future access for maintenance should be considered at the design stage. Windows should be of a type that can be cleaned from within the building.
Inadequate trimming of dormer into main roof	The trimmer and its supporting members have to carry the truncated portion of the main roof resulting from the dormer. The design may impose additional loads on the floor joists. Calculations to ensure the floor joists can carry this additional load should be carried out by a Structural Engineer.

HAPM GUIDANCE

The design drawings should show that the consequences of installing dormers have been fully considered and that there are adequate facilities for access and maintenance. The type of roof will be noted, ie warm, cold, inverted or hybrid. In all cases, the Technical Audit Unit will require a copy of the roof and trimmer beam design and details.

Details

COMMON DEFECTS	CRITERIA/GUIDANCE
Water penetration at junction of dormer with main roof	The junction of the dormer with the main roof is very vulnerable to rain penetration. Flashings should be taken far enough up the main roof to exclude wind-driven rain and snow. Most specifications require the height of the projected flashing to be 150mm which means the length of the flashing should be: • 300mm up the slope for 30° pitch • 355mm for 25° pitch • 580mm for 15° pitch and so on, increasing for flatter pitches.
Water penetration at partially inset dormers	Flashings at the point where the dormer changes from 'projecting' to 'inset' should be detailed carefully.
Water penetration via back of window frame	Most framed dormers will incorporate a lightweight, vertical cladding which has to be joined to the window frame(s). Good detailing practice is essential and at least 'two lines of defence' against water penetration should be incorporated into the design.
Water penetration via corner of dormer	The dormer cladding will inevitably have to include an external angle which should be fully detailed.
Water penetration under apron flashing at base of dormer	Wind-driven rain can penetrate via the underside of an apron flashing under the window cill. Flashings should be fully detailed. The projected height of the flashing should be 150mm.
Ceiling insulation blocks eaves ventilators	Ventilation requirements should be clearly detailed so that they are not omitted during construction nor blocked by any insulation. The clear depth of ventilation void should be 50mm. Refer to BRE *Defect Action Sheet 1* and *4*.
Condensation at cold bridges	Overlaps of insulation from the main roof to the dormer roof to the dormer cheeks should be specified. The paths of greatest heat loss should provide a U value not greater than 1.2W/m²degC.
Inadequate protection to dormer cheeks and windows from dormer roof	The roof over a dormer may include the eaves, verge and gable configurations of the main roof but to a smaller scale. These details should give good protection to the dormer walls and windows below.

HAPM GUIDANCE

Time and attention should be given to architectural details of dormers. Adequate accessibility to parts requiring routine maintenance should be provided. Good detailing practice should be applied. Junctions with roof at top, sides and bottom should be clear. Where flashings are complex, they should be considered in all three dimensions.

Materials

COMMON DEFECTS	CRITERIA/GUIDANCE
Inadequate durability of cladding to dormer cheeks	The claddings for dormer cheeks, and their fixings, should be as durable as for any other external wall. Where plywood is used, it should be marine-quality to BS 1088: 1966 *Specifications for plywood for marine craft.*
Inadequate durability of timber in dormer	All timber, including tiling battens, fascias, etc should be treated with an organic solvent fungicide and/or insecticide applied by a double-vacuum process. The treatment should leave a tinted finish so that untreated timber can be easily distinguished on site.
Inadequate durability of windows in dormer	Since in an average dwelling, dormer windows are normally more exposed and less accessible than other windows, aluminium, uPVC or hardwood frames should be considered, even though these may not have been used elsewhere in the building.
Inadequate durability of roof covering	Access for maintenance is often difficult, therefore a long-life covering with adequate solar protection should be specified. See Topic 4.2 for more detailed advice.
Flashings become brittle and split	Cheap materials tend to be less durable. Lead flashings should be Code 4 for small details and Code 5 for large details and dormer roofs. Proprietary flashing materials should not be used unless they are either supported by a BBA Agrément Certificate or conform to the relevant British Standards.
Mesh for ventilators too small or too large	The mesh for ventilators should be small enough to prevent the entry of birds and insects but large enough not become blocked by debris, corrosion or paint. Refer to BRE *Defect Action Sheet 1* and *4*.
Vapour control layers too flimsy or too permeable	Vapour control layers should be resistant to perforation and to the passage of water vapour.

HAPM GUIDANCE

All materials should be compatible with each other and to the relevant British Standards or should be supported by a BBA Agrément Certificate. Bearing in mind the inaccessibility of dormer roofs, materials should be selected for their durability and should have a life extending beyond the expiry date of the Major Defects Insurance Policy. See Topics 4.1 and 4.2. Refer also to *Lead Sheet in Building* published by the Lead Development Association.

Workmanship

COMMON DEFECTS	CRITERIA/GUIDANCE
Flashing upstands too shallow	Projected heights of flashing upstands should not be less than 150mm, measured vertically (not up the slope). Weathertightness is essential since persistent rain penetration may be concealed within the construction of a dormer leading to major damage or decay before the penetration becomes apparent.
Lead flashings used in too long lengths	Lead flashing should not be laid in too long a length because the lead pulls and splits under thermal movement. The recommended limit is 1.5m. Refer to *Lead Sheet in Building* published by the Lead Development Association for guidance.
Membranes not fixed down	Wind uplift on the flat roofs of dormers may be accentuated by the wind-flow pattern over the main roof. Flat roof membranes should be fixed securely at all edges.
Perforations in vapour control layers	All edges in the vapour control layers should be sealed. Holes formed for vent pipes, flues, handrail standards, etc should be sealed carefully. Any tears or accidental damage should be made good to achieve a high vapour resistance.
Moisture-sensitive decking saturated during construction	Particular care should be taken to prevent moisture getting trapped in a wood wool or timber deck during construction otherwise the deck elements can swell. When this trapped moisture dries out, the decking sheets shrink at the end joints. Over time, this cyclical thermal expansion and contraction will cause fatigue and splits in the felt.
No preservative treatment to cut ends of timber members	Timber framed dormer roofs necessarily require timber to be cut to length at site. Ends of such members should receive a preservative treatment applied by brush and equivalent to the specified treatment to suit the species of wood being used. Similar treatment is required when the horns of the frames are cut to length at site.
Felt laid on flat roofed dormers in cold, wet or windy weather	In cold weather, rolls of bitumenized felt should be warmed before they are unrolled or they will tend to crack (hence the lower limit of 5°C in most specifications for work in cold weather). The general problems of laying bitumenized membranes in cold weather can be overcome if heat and shelter are provided to the work area. Moisture prevents the adhesion of bitumen to any surface. Refer to BRE *Defect Action Sheet 63*.

HAPM GUIDANCE

The size and shape of dormer windows vary considerably and, therefore, so does the work involved in preparing the flashings and making the arrangement watertight. Roof coverings and their substrates should be laid in accordance with the manufacturer's instructions and the relevant code of practice for the particular material, for instance, CP 144: Part 3: 1970 *Roof coverings: Built-up bitumen felt.* Builders and Supervisors should pay particular attention to the above common defects and to construction quality at all details. The Technical Audit Unit Site Inspectors will undertake a 'quality count'.

Design

COMMON DEFECTS	CRITERIA/GUIDANCE
Insufficient cross-sectional area of flue	The flue should have a cross-sectional area not less than the boiler outlet. Refer to Appendix A in BS 5440: Part 1: 1990 *Installation of flues and ventilation for gas appliances of rated input not exceeding 60kW (1st, 2nd and 3rd family gases): Specification for installation of flues* for guidance on the sizing of the flue.
Flue has sharp turns, bends and pockets	Where possible, flues should be vertical. Bends should not exceed: • 30° from the vertical for solid fuel, domestic-sized appliances • 45° for gas or oil appliances. Refer to Section 2 in *Approved Document J1/2/3.*
Insufficient height of flue to solid fuel or oil appliance	To ensure a satisfactory clearance of the products of combustion, the chimney should normally terminate: • 600mm above ridge level, or • 1000mm above roof slope level. Refer to *Approved Document J.* In bungalows, a sufficient height of flue should be provided; generally, 3 metres is satisfactory. The height of the chimney should be considered in conjunction with the height of other buildings in the vicinity. The slenderness ratio of the stack should be checked. Refer to Appendix A in BS 5440: Part 1: 1990.
Flues terminate close to adjacent gable walls	At steps or staggers in rows of terraced houses, flues should not terminate alongside gable walls but should extend to 600mm above the adjacent verge level.
Reduction of acoustic performance of separating walls by flues	Precast concrete flues should be designed or located so that the acoustic insulation of any separating wall is not compromised. Flues should not be located back-to-back. Flue blocks should be bonded in with the wall blocks.
Chimney does not 'draw' properly	When the chimney does not 'draw' properly, it may be due to the incorrect size or shape of the throat or fireplace. Where the chimney opening is too high, the supply of air may be drawn above the fire leaving it deprived of oxygen. Modern 'sealed' houses should have permanent ventilation to rooms containing solid fuel appliances, unless the fire has a special dedicated air supply.
Poor siting of balanced flue terminals to gas appliances	Balanced flue terminals should be sited where there is a free flow of fresh air and where they are unlikely to be damaged by opening doors or windows, or by vandalism. Access ways, lightwells or carports should be avoided. For general guidance, refer to the British Gas publication *Gas in Housing.*

Design

COMMON DEFECTS	CRITERIA/GUIDANCE
Flues in timber framed dwellings made up on site	The flue should be a separate self-supporting structure or it should be able to accommodate movement and shrinkage of the timber frame. Preferably, the flue should be supplied as a complete kit from one manufacturer with written instructions for its assembly. Refer to the British Gas publication *DM2*, September 1984, for detailed guidance on flues and timber frame construction.
No access provided for cleaning chimney	Since chimneys for oil-fired or solid fuel appliances should be swept regularly, a cleaning door for access for brush cleaning should be provided near the base of the chimney and at all bends.

HAPM GUIDANCE

The design of chimney and flue should be shown to comply with *Approved Document J1/2/3* and relevant parts of *Approved Documents A, F, L* and *M.* Refer to BS 5440: Part 1: 1990, BS 5449: Part 1: 1990 *Specification for forced circulation hot water central heating systems for domestic premises*, BS 6461: Part 1: 1984 *Installation of chimneys and flues for domestic appliances burning solid fuel (including wood and peat): Code of practice for masonry chimneys and flue pipes*, and BS 6461: Part 2: 1984 *Installation of chimneys and flues for domestic appliances burning solid fuel (including wood and peat): Code of practice for factory-made insulated chimneys for internal applications.* Reference may also be made to publications by the Solid Fuel Advisory Service and the Real Fire Heating Association.

Details

COMMON DEFECTS	CRITERIA/GUIDANCE
Water penetration around flue	Where a flue passes through a tile or slate roof, the waterproofing arrangement should consist of a purpose-made plate, with a minimum upstand of 150mm above the highest point of its junction with the roof. Where a flue passes through an asphalt or other flat roof, the waterproofing arrangement should be fully detailed. The flue should be isolated by a sleeve fitted securely to the roofing material.
Flue pipes placed too close to combustible materials	The distances at which flue pipes should be placed from combustible materials are shown in Diagram 2 in *Approved Document J1/2/3.* Particular care should be taken where single-walled flue pipes pass through floors or into roof spaces. Sleeves should be considered to provide sufficient clearance between the flue pipework and any combustible materials. Refer to BRE *Defect Action Sheet 60.*
Discontinuity of cavity when chimney constructed in external wall	Walls should be detailed so that the cavity is continuous around the chimney stack.
Inadequate detailing of dpcs in brick chimney stacks	Dpcs should be provided at the higher junction of the chimney with the roof. On steeply-pitched roofs, a second dpc may also be required at the lower junction.
Inadequate resistance of chimneys to sulphate attack	Condensation arising from water vapour in the flue gases may contain sulphates. In unlined chimneys, this condensation could penetrate into the mortar causing expansion and resulting in cracking, bending or twisting of the stack. Ideally, the chimney should be built using liners fitted correctly. Sulphate-resisting cement should always be used for chimney brickwork mortar.
Chimney walls too thin	Brick or block chimney walls should be at least 100mm thick: • between one flue and another • between a flue and the outside air • between a flue and another part of the dwelling. Flue walls should be 200mm thick: • between a flue and another compartment of the building • between a flue and another building or dwelling. Refer to Clause 2.14 in *Approved Document J1/2/3.* Where hollow or cellular blocks are used in the chimney, the voids should be filled with concrete otherwise the inner walls and ribs will break up due to the heat.
Movement and cracking of plaster finish due to overheating of flue blocks	To prevent overheating of the flue blocks, insulation and plaster reinforcement may be required over the block. Refer to the flue block manufacturers' recommendations.

HAPM GUIDANCE

Masonry or precast chimneys and flues should be detailed and show the arrangement at roof, intermediate and ground floor intersections.

Materials

COMMON DEFECTS	CRITERIA/GUIDANCE
Incorrect type of flue terminal fitted	Flues should terminate with an outlet appropriate to the type of fuel being used. Refer to BS 5440: Part 1: 1990 *Installation of flues and ventilation for gas appliances of rated input not exceeding 60kW (1st, 2nd and 3rd family gases): Specification for installation of flues* and BS 6461: Part 1: 1984 *Installation of chimneys and flues for domestic appliances burning solid fuel (including wood and peat): Code of practice for masonry chimneys and flue pipes.*
Uninsulated flue pipes used externally	Single-skin, uninsulated flue pipes should not be used externally, especially for solid fuel or oil-fired boilers since the updraught will be poor if the flue temperature is too low and this results in poor combustion. Factory-made insulated chimneys should comply with the requirements of BS 4543: Part 2: 1976 *Factory-made insulated chimneys: Specification for chimneys for solid fuel appliances* and BS 4543: Part 3: 1976 *Factory-made insulated chimneys: Specification for chimneys for oil fired appliances.*
Chimneys not lined properly	Brick or other masonry chimneys should be lined with a suitable durable lining. Chimneys for solid fuel appliances should be able to withstand a temperature of 1100°C without undergoing any change which would impair the stability or performance of the chimney.
Flue blocks do not bond in with wall blockwork	Flue blocks should be made from refractory materials, such as kiln-burnt aggregate and high-alumina cement. Preferably, the flue blocks should be obtained from the same manufacturer as the general blockwork since the tolerances are then likely to be compatible. Note that flue blocks are more resistant to airflow than flue pipes.
Flue not designed to resist heat loss or corrosion	The chimney should be lined with insulating and corrosion- resistant material to reduce heat loss and to reduce the risk of damage to the chimney fabric due to condensation. Where the flue temperature is too low, the updraught will be poor resulting in poor combustion and smoke and gases blowing back into the dwelling. Refer to BRE *Defect Action Sheet 126.*
Corrosion of lead trays, aprons and flashings	In the presence of moisture, free lime in new mortar or concrete can initiate a slow corrosive attack on lead. Lead sheet in contact with new mortar or concrete should be protected by a coating of bitumen paint. External flashings tucked into brickwork joints do not need protection.

HAPM GUIDANCE

Builders or Contractors should hand over to the tenant the manufacturer's operating instructions for all appliances. Refer to BS 8308: 1986 *Code of practice for installation of domestic heating and cooking appliances.* Components of chimneys and flues will be expected to have a life in excess of the Major Defects Insurance Policy, except that, normally, domestic balanced flues and fan-assisted flues will be replaced with the boiler.

Workmanship

COMMON DEFECTS	CRITERIA/GUIDANCE
Flue pipe joints wrong way up	The socket end of a pipe should always be uppermost. However, note that twin-walled flue pipes should be installed with the spigot end uppermost.
Flue pipes jointed in floor depth	Flues should be assembled so that joints in the pipes do not occur in the floor depth.
Annular space between liner and chimney not sealed	The annular space between the liner and chimney should be fully sealed at both the upper and lower ends so that the only openings are: • the opening for the appliance • the opening for the terminal outlet.
Joint between flue system and terminal not gastight	Where the flue system includes a ridge terminal, care is required to ensure that the ridge terminal has been sealed to the flue pipe.
Twin-walled flue pipes and fittings modified at site	Twin-walled flue pipes should not be cut to length at site, unless specifically permitted by the manufacturer's instructions.
Adjustable bends in flue pipes strained beyond practicable limits	When the bend is overstressed, there is likely to be a breakdown of the inner wall of the pipe. All pipe sections should be examined before assembly and any section with damaged joints or internal damage should be discarded.
Masonry chimneys not cored during construction	Traditionally, a 'core' consists of a sack attached to a rope and filled with resilient material to fit the flue. The core is pulled up by the rope as the work proceeds and while the mortar is still moist. The chimney should be swept on completion.
Flue pipe casings made from unsuitable materials	Casings for flue pipes should be made from non combustible materials, with clearance to the flue pipes as shown in Diagram 2 of *Approved Document J1/2/3*.
Dpcs and flashings in chimneys omitted or positioned incorrectly	Dpcs and flashings to chimneys should be installed strictly in accordance with the drawings. Refer to BRE *Defect Action Sheet 94* and *95*.
Inadequate jointing of masonry and flue liners	Unfilled mortar joints are not acceptable. On completion, the following checks should be made: • the flue diameter, to ensure it is as specified • the refractory lining, to ensure it is as specified • the bore, to ensure it is smooth and clean, and free from protruding rings of mortar. Refer to BRE *Defect Action Sheet 127*.
Structural timber roof members cut away to clear path for flue	No structural timber roof member, including noggins, packings and bracing, should be cut.
Flues located too close to combustible materials	Combustible materials include timber, roofing felts, moisture or vapour control layers and some insulants. Refer to BRE *Defect Action Sheet 60*, but note the clearance of the flue to combustible material as stated in *Approved Document J1/2/3*.

Workmanship

COMMON DEFECTS	CRITERIA/GUIDANCE
Inadequate support to flue pipes	Flue pipes should be supported at centres not greater than 1.8m, and at all changes of direction. Refer to BRE *Defect Action Sheet 60.*

HAPM GUIDANCE

Builders or Contractors should test each flue for efficiency. The Gas Board offer an inspection service for gas appliances which also includes other safety checks.

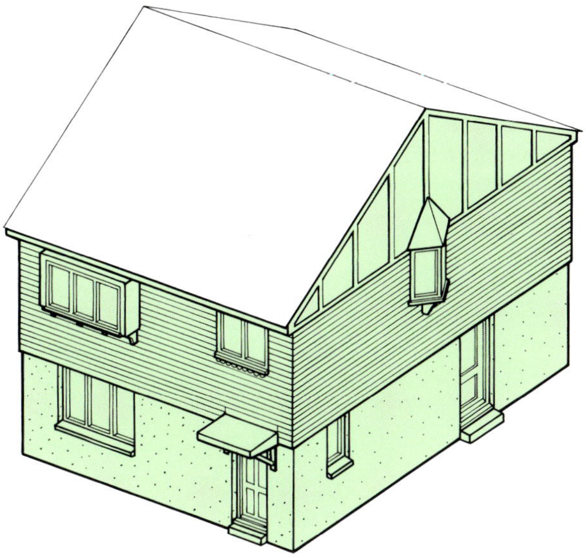

HAPM

HOUSING ASSOCIATION PROPERTY MUTUAL LIMITED

WALL CLADDINGS

PURPOSE-MADE JOINERY

BAYS, PORCHES, CANOPIES, ORIELS

EXTERNAL BALCONIES, STAIRS, RAMPS

5

Design

COMMON DEFECTS	CRITERIA/GUIDANCE
Shrinkage cracks in render aesthetically unsatisfactory	Formal movement joints should be provided in render. Refer to BS 5262: 1976 *Code of practice. External rendered finishes* for spacing requirements. Rough-textured finishes should be considered as they tend to hide local inconsistencies in finish and so weather better and are less prone to cracking.
Render specified where not appropriate	Render should not be specified for walls exposed on both faces, for instance parapets, since any moisture trapped in the wall could cause problems, such as debonding of the render.
Time not allowed in contract programme for rendering operations	Rendering may not be suited to 'fast track' construction programmes as brick or block walls need to consolidate and shrink before the render is applied. Seven days should be allowed between successive coats of render.
Insufficient resistance of cladding to impact	Low-level vertical tiling is vulnerable to accidental damage and vandalism so its use should be restricted to levels above the ground floor. Refer to BS 8200: 1985 *Code of practice for design of non-loadbearing external vertical enclosures of buildings* for guidance on the selection of claddings in terms of impact resistance.
Insufficient resistance of cladding to wind suction	Manufacturers will usually offer guidance on the amount of primary fixing necessary and will relate this to the exposure (wind suction) conditions. Secondary fixings, ie batten fixings into lightweight blocks, should be carefully considered and the required frequency calculated. The bottom edges of slates may require fixing back. Refer to BS 5534: Part 2: 1986 *Slating and tiling: Design charts for fixing roof slating and tiling against wind uplift.*
Insufficient resistance of cladding to rain penetration	The manufacturer should confirm the suitability of cladding for 'moderate' and 'severe' exposure situations.
Condensation on back of cladding or on back of membrane	Any sheathing membrane behind the cladding should be vapour permeable if it is on the cold side of the insulation (not less than 36g/m² to BS 3177: 1959 *Method for determining the permeability to water vapour of flexible sheet materials used for packaging*). If the cladding itself is impermeable, provision for ventilation behind the cladding and/or drainage of the condensate should be made.
Limitations of plastics or timber weatherboard in fire	Plastics or timber weatherboard do not achieve Class '0' surface spread of flame and cannot be used within one metre of a boundary. When they are used in excess of one metre from a boundary, they affect the calculation of 'unprotected area'. Refer to *Approved Document B2/3/4.*
Inadequate specification of surface preparation for rendered finishes	For render on brickwork (if keyed bricks are not used), the brickwork joints should be raked out to a depth of 15mm to provide a good key. Dense smooth surfaces may require bush-hammering orgrit-blasting to provide a good key, or the use of a bonding agent may be considered.

Design

COMMON DEFECTS	CRITERIA/GUIDANCE
Extension of fire spread and damage in flats and maisonettes over two storeys high where combustible thermoplastic insulants have been used in external wall	Horizontal fire barriers should be provided to subdivide combustible insulants at not more than two storey intervals to prevent the spread of fire upwards. Reliance should not be placed on plastics fixings for insulation which should be supplemented by stainless steel or non-ferrous fixings. Refer to BRE *Defect Action Sheet 131* and *132*.

HAPM GUIDANCE

All proprietary claddings should have a proven record for exposure and durability and should be of a British Standard quality or supported by a BBA Agrément Certificate. Full details of fixings should be shown.

Details

COMMON DEFECTS	CRITERIA/GUIDANCE
Absence of overhang protection	Details should be designed to shed water clear of the wall if streaking is to be avoided. When rendering, bell-mouth drips should be provided over openings and at dpcs. Render should never bridge a dpc. Refer to BS 5262: 1976 *Code of practice. External rendered finishes* for details of sills, string courses, parapets, etc.
Inadequate detailing of cladding around openings	Very informal and inadequate detailing is often used around windows set in tile hanging. Most cladding manufacturers offer standard details. 'Straight through joints' and total reliance on mastic should be avoided.
Inadequate protection of cladding from condensation and water absorption	Timber, plastics and metal cladding should be provided with ventilation and drainage to the rear face. Timber cladding should receive the first coat of paint or stain to both faces and all edges before erection.
Moisture penetration to edges of timber cladding	The end grain of timber should not be in contact with brickwork. Details should show the bottoms of vertical boards cut to chamfer with free outward drainage over the flashing below. All ply edges should be sealed. Helpful details are included in *Exterior Cladding of Redwood and Whitewood* published by The Swedish Finnish Timber Council.
Continuity of render interrupted by decorative 'Tudor' boarding	Timber should be fixed to the wall only after the first coat of render has been applied. The second coat of render is then placed between the timbers, which should be treated with preservative.
Cladding components become detached in fire	Components should be detailed so that they do not become detached by heat and, if they do fall, they should not represent an unacceptable hazard to occupants escaping from the building or to fire fighters.
No formal joint between cladding and vertical abutment	Generally, the vertical junctions of cladding with masonry or other elements should include a drained joint, a lap joint or a formal seal. All joints should be dimensioned to accommodate the extremes of thermal movement likely in the cladding. Mastic joint design should take account of the movement capability of the mastic specified. Helpful information on movement and joint design is included in *Manual of good practice in sealant application* by The Sealant Manufacturers' Conference.
Cracking of render at movement joints or changes in background material	At movement joints, the render should stop, or be Vee-jointed and sealed. Where the background material changes, the change should be bridged with breather paper (to prevent adhesion) behind metal lathing. The lathing should be fixed with a gap behind it so that the render can 'key' round the mesh.

Details

COMMON DEFECTS	CRITERIA/GUIDANCE
Weathering and discolouration of stone masonry cladding	Weathering and discolouration of stone masonry cladding may be due to: • an uneven flow of water over the surface • breakdown of limestone surfaces due to crystallisation of salts • frost damage in fine porous limestone • incorrect bedding • attack by chemical pollution. Details should be designed to shed water and keep walls dry.

HAPM GUIDANCE

Scheme drawings should show details of the head and base of all claddings, corners, junctions with other claddings, junctions with window/door frames and junctions with adjoining construction.

Materials

COMMON DEFECTS	CRITERIA/GUIDANCE
Incorrect mix for render	Care should be taken when mixing render because: • if the mix is too weak, it may not act as an effective barrier to rain penetration • if the mix is too rich or too wet, it may shrink and crack leading to rain penetration • if the mix contains too high a proportion of fine sand, it will crack. The first coat of any rendering should be made with sulphate-resisting cement to accommodate any possible presence of sulphates in the brickwork. Refer to BS 5262: 1976 *Code of practice. External rendered finishes*, BRE *Digest 196* and BRE *Defect Action Sheet 38*. Sand should be coarse and comply with Table I of BS 1200: 1976 *Specification for building sands from natural sources*. The mix design should be checked with the relevant brick or block manufacturer. The colour pigment should be limited to 10% of the weight of cement and be only to BS 1014: 1975 *Specification for pigments for Portland cement and Portland cement products*.
Incorrect number/thickness of render coats	The number and thickness of render coats should be specified to BS 5262: 1976. As a rule, no coat should be stronger or thicker than the preceding coat. Note that metal lathing requires three coats of render.
Corrosion of metal render strips and beads	Internal-quality (plastering) beads are often used in error. Although most manufacturers make special external render-quality galvanized render stops and beads, stainless steel is preferred.
Inadequate durability of metal trims and fixings	Pressed metal trims for use with some tiling systems may be only of galvanized steel and would require additional site protection and regular recoating. Marine and industrial environments are particular risk locations. All steel nails should be plated or galvanized. Non-ferrous nails are preferred for tile hanging except that aluminium nails should not be used in CCA-treated timber.
Inadequate durability of timber battens	The following should be double-vacuum or pressure-treated with preservative: • all battens • all counter battens • any make-up pieces of timber, spacers or noggins that are used to supported a cladding.
Corrosion of dissimilar metals	Materials should be selected so that there is no adverse chemical or electrolytic action between them.
Inadequate durability of tiles, slates, fixings	Cement-based artificial slates will deteriorate in acid environments. Fixing clips should be stainless steel.
Inadequate long-term durability of GRC or plastics items	Fading, staining and long-term embrittlement may become a problem.
Failure of protective/decorative coating to profiled steel cladding	The protective coating or finish may be unsuitable for the exposure conditions, and may blister and become detached, or may degrade in sunlight. Chafing at joints and fixings can occur due to thermal movement. This can lead to the erosion of the coating and the exposure and subsequent corrosion of the metal.

Materials

COMMON DEFECTS	CRITERIA/GUIDANCE
Incorrect specification for applied finishes	For rendered surfaces, only paints of high vapour permeability should be used otherwise there will be the possibility of sulphate attack or frost damage.

HAPM GUIDANCE

Materials should comply with the requirements of a British Standard or be supported by a BBA Agrément Certificate. Proprietary renders and cladding systems should be supported by a BBA Agrément Certificate. Treatment certificates should be required by Supervisors for preservative-treated timber.

Workmanship

COMMON DEFECTS	CRITERIA/GUIDANCE
Render applied to wet or contaminated surfaces	Surfaces should be dry and clean. Efflorescence should be removed by brushing the surface.
Confusion on site between internal and external quality plaster trims	Stainless steel or external-quality galvanized lathing and corner beads should be used. Internal-quality plaster beads should not be used externally.
Departure from specified mixes of render	If premixed render cannot be used, the ingredients of the mix should be carefully measured at site by means of gauge boxes.
Map pattern cracking of render	Map pattern cracking of render can result from the shrinkage of too wet a mix. The mix is often too wet because the sand used is too fine.
Cracking or curling of render	Cracking or curling of render can occur because successive coats are not weaker or thinner than the first coat. At least seven days should be allowed between successive coats of render so that the later coats can fill shrinkage cracks in the first coats and so that shrinkage of the first coats will not put stress into the subsequent coats. Refer to BS 5262: 1976 *Code of practice. External rendered finishes*, BRE *Digest 196* and BRE *Defect Action Sheet 37*.
Inadequate fixing of tile cladding or weatherboarding to structure	The security of the batten fixings should be checked. Note that there is a particular risk with nailed fixings into lightweight blocks.
Detachment or displacement of tiles or slates at corners, edges and abutments	Small, irregular cut tiles are difficult to fix properly. The work should be set out so that small irregular tiles are avoided. Alternatively, tile-and-a-half specials should be used.
Buckling of plastics cladding	Cladding should not be fixed during extremes of temperature. Joint coupling parts should be used if possible. Each plank of cladding should be nailed from the centre. Butt joints should be staggered. Clearance around fixings may be required to accommodate movement.
Cracking or splitting of cement-based cladding sheets	Sheeting is liable to crack or split when thermal movement occurs if insufficient clearance has been provided around fixings, or the fixings are too tight. Refer to BS 5247: Part 14: 1975 *Code of practice for sheet roof and wall coverings: Corrugated asbestos-cement* and to manufacturers' recommendations.
Insufficient overlap of timber weatherboarding planks	Planks should have at least 10mm of overlap. Tongued and grooved boards should be fixed with the tongue upward.
Incorrect nails used in fixing weatherboarding	Galvanized or non-ferrous nails should be used. Aluminium nails should not be used in CCA-treated timber. The length of nails should be not less than two-and-one-half times the board thickness.
No preservative treatment to timber cut at site	All ends cut at site, notches, etc should receive a preservative treatment applied by brush in two coats, which is equivalent to the original preservative treatment.

Workmanship

COMMON DEFECTS	CRITERIA/GUIDANCE
Paint system applied to damp surfaces	Most paint failures are caused by unsuitable conditions during application.

HAPM GUIDANCE

The Technical Audit Unit Site Inspectors will undertake a 'quality count'. Supervisors should retain timber preservation certificates for inspection. Supervisors should ensure that manufacturers' fixing instructions are also retained on site for reference.

Design

COMMON DEFECTS	CRITERIA/GUIDANCE
Inadequate or incorrect specification of timber species	The terms 'softwood' or 'hardwood' are not sufficient: the species should always be specified by name Appendices B and C in BS 1186: Part 1: 1986 *Timber for and workmanship in joinery: Specification for timber* list species of softwood and hardwood timbers and their suitability for use in various joinery situations. Some species, eg European Whitewood, are suitable externally only if they are treated with preservative.
Inadequate specification for timber quality	Timber quality should be specified by reference to a 'Class', which is defined by the minimum acceptable level of defects. Defects include knots, splits, resin pockets, sapwood, widely-spaced growth rings, etc. BS 1186: Part 1: 1986 defines four classes of timber quality.
Method of jointing not specified	The method of jointing timber members should be specified. For external doors, mortice and tenon joints are preferred to dowel or comb joints. Window casements and smaller timber sections may be comb-jointed.
Inadequate specification for glue type	All joints should be glued. Glues for exterior use should be weather-resistant and should have gap-filling properties. The two basic glue types are: • synthetic formaldehyde resins of urea, melamine, phenol and resorcinol • PVA emulsions. However, within these two groups, the glues have widely varying properties. BRE *Digest 340* reviews the properties of each type of glue and provides guidance on choosing the correct adhesive. It also covers the difficulties of glueing timbers impregnated with preservatives and/or fire retardants. BRE, Princes Risborough Laboratory *Technical Note No 31* provides more detailed advice on glueing preservative-treated wood.
Purpose-designed joinery components not strong enough	The design of joinery items, such as full-height glazed screens, balustrades and staircases, should be checked by a Structural Engineer to ensure they are capable of withstanding wind loads and other imposed loads.

HAPM GUIDANCE

There is very little helpful guidance on the design of joinery. Typically, designers prepare only outline drawings and rely on the skill of joinery manufacturers. Consequently, many joinery items arrive on site held together by rusting nails and adhesives that are not adequate for outside use. Unfortunately, true 'WBP' adhesives are difficult to use and timber preservation is an unwelcome interruption in the manufacturing process (necessary after machining but before assembly). The potential for failure is further increased by the problem of adequately seasoning timber to avoid shrinkage and movement in service and the problem of accurate species recognition. Designers should provide comprehensive specifications to avoid premature failure and inspect joinery works during fabrication.

Details

COMMON DEFECTS	CRITERIA/GUIDANCE
Inadequate weathering of screens/windows	Every opportunity should be taken to shed water clear of the construction by using cills and drips that have been weathered properly and are placed appropriately. Drips are of most value when placed below large areas of impervious material, such as glass or cladding panels. Where drips are planted, they should be either bedded on a sealant strip and fixed with screws, or housed into a rebate. *Principles of Simple Weathersealed Timber Window Design* published by The Swedish Finnish Timber Council gives some good practical guidance.
Inadequate weathering of horizontal members	External surfaces of horizontal members should be weathered with a slope of about 1 : 8. Internal horizontal surfaces need not be weathered except below glazing to provide a run-off for condensation, but it is better to provide a gutter or channel which drains to the outside.
Clearances between openings and fixed parts not properly considered	The clearance between the meeting parts of window casements and door leafs should be sufficient to allow for a paint build-up and to accommodate any weatherstripping without distortion, for example: • a 3-4mm gap should be left where there is weatherstripping • a 2mm gap should be left in other cases.
Glazing beads mitred in lower corners	Mitring of the glazing beads in the lower corners will lead to water penetration to the end grain of the bead. This increases the risk of rotting. Side beads should be scribed to the bottom bead, leaving a gap of 3mm. The bottom bead should extend the full width of the opening and be properly weathered. It is better to use aluminium or stainless steel beads for the bottom of the window.
Insufficient hinges	External doors and windows exceeding 1.5m in height should be provided with 3 hinges.
Waterbars incorrectly positioned on inward opening doors	Waterbars should be located about 10mm inside the face of the doorstop so that water driven into the gap between the door and the stop is directed to the outside at the bottom. The bottom edge of the door should be rebated over the waterbar and a weather moulding should be provided.
Leaking of internally-beaded glazing	For security or maintenance reasons, it is sometimes necessary to specify internally-beaded glazing. However, internally-beaded glazing is prone to leakage because any water that passes the glazing compound is effectively trapped on the inside of the glazing rebate. Only the best quality glazing systems will work in this situation. Refer to *The Glass and Glazing Federation Handbook* for guidance.
No drips provided on projecting cills	Where a horizontal surface occurs under a projecting member, water will tend to be blown back by the wind into the horizontal joint. To prevent this, a 'drip' or trough at least 8mm wide and 5mm deep should be specified.
Timber cills too thin	Timber cills are prone to curl if they become too thin and/or too wide. Note that the slope of the cill reduces the thickness and a drip under the front edge reduces the effective thickness still further.

Details

COMMON DEFECTS	CRITERIA/GUIDANCE
Rounded arrises not specified	Since paint films tend to pull away from sharp arrises resulting in a weakness which will lead to early paint failure, arrises should be specified as rounded.

HAPM GUIDANCE

All timber sections should be drawn to a large scale and dimensioned. Types of joint and glue should be specified.

Materials

COMMON DEFECTS	CRITERIA/GUIDANCE
Component sizes too large to be easily handled	Some components, such as staircases, have to be broken down into sections so that they can be fitted into position on site. These large components should be designed to be broken down into easily handled sections so that manufacture can be carried out in an off-site workshop. The sections should be small enough to go through doors. Joints in components should be properly designed to be expressed or concealed and each half should be structurally independent.
Prefinished components damaged when installed too early	When prefinished components are installed too early in the construction process, they can be damaged by following trades. Components and their surroundings should be designed so that they can be installed as late as possible in the construction process. Dummy frames and formers may be necessary to ensure that openings are of the required size.
Non-hardening glazing compounds harden	Glazing compounds should be carefully selected for both internally and externally beaded glazing. Mastics generally perform better than glazing compounds.
Corrosion of mild steel fastenings used in external joinery	Only heavily-plated or non-ferrous fastenings should be used in external joinery.
Inadequate durability of weatherstrips	Weatherstrips often fail early in the life of joinery. Since their replacement will be considered a common maintenance operation, consideration should be given to ease of replacement.
No preservative treatment to non-durable timber	Normally, organic solvent preservatives applied by a double-vacuum process are to be preferred. The treatment cycle should be selected to suit the timber species as overtreatment can lead to premature paint failure.

HAPM GUIDANCE

All materials, including timber specials, glues, mechanical fixings, glazing compounds, mastics and preservative treatments, should be specified fully.

Workmanship

COMMON DEFECTS	CRITERIA/GUIDANCE
Joinery components badly stored on site	Joinery components should be stored under cover, preferably in a completed part of the building. Where this is not possible, they should be covered with a tarpaulin on a level base of timber bearers to allow air circulation and the drainage of trapped water.
Timber with excessively high moisture content used	Ideally, internal joinery should not be introduced into new buildings before the plaster and brickwork has had a chance to thoroughly dry out naturally, which can take up to 3 months. Where excessively moist timber is used for, say, tongued and grooved boarding, the shrinkage across the grain may be so great as to result in the tongues being pulled out of the grooves. Timber should be dried to a moisture content of about 14-15% before making up.
Joint meeting surfaces not closely fitted	The exposed meeting edges of framed joints should be flush and in tight contact. BS 1186: Part 2: 1988 *Timber for and workmanship in joinery: Specification for workmanship* allows for a tolerance of 0.5mm on the unexposed joint faces. A greater tolerance of 1mm is allowed where the flush meeting surfaces are stepped, grooved or relieved at the meeting face.
Panels not given sufficient clearance to accommodate movement	Plywood or other timber panels in framed joinery should not be glued. They should be allowed to move freely to accommodate moisture movement. BS 1186: Part 2: 1988 recommends a clearance of 3mm. When using tongued and grooved boarding, each board should be single-fixed to accommodate movement.
Surface of timber not sufficiently prepared to receive paint or varnish finish	Machining of timber should take place at a moisture content close to that of its moisture content in service. For paint finishes, timber should be knotted before priming. Arrises should be removed and the whole component completely sanded between coats. If necessary, machine marks should be removed by hand finishing. All defects should be filled with either an internal- or external- quality filler, as appropriate. When using a clear finish, the ability of any filler to take the finish should be checked beforehand.
Inadequate bracing of pre-assembled items for transport and handling	To avoid distortion or racking of joints, temporary bracing should be used during transportation and handling.

HAPM GUIDANCE

The Technical Audit Unit Site Inspectors will be looking for adequate storage arrangements for joinery.

Design

COMMON DEFECTS	CRITERIA/GUIDANCE
Foundation less supportive than that to main structure	Ground-supported appendages should have the same depth of foundation as the main building. Timber columns need adequately designed shoes at ground level.
Cantilevered structures inadequately engineered	Design calculations should take account of flat surfaces being used for access and maintenance. Cantilevered appendages should be anchored very firmly.
Window frames inadvertently used for structural purposes	Many cantilevered appendages take advantage of the strength in the window frame. Load paths should be properly designed and soft joints should be used to avoid stressing the non-loadbearing elements and to allow their easy removal.
Poor rainwater shedding	Flat roofs of bays, porches, canopies and oriels should not drain towards a wall. Generally, rainwater drainage should be provided, unless the roof area is very small.
Condensation on surfaces	Enclosed porches, bays and oriels are all susceptible to condensation. The concrete bases to oriels are potential cold bridges and therefore should be insulated.
Inadequate resistance to fire	Projections from the face of the building may constitute an external wall and therefore may require fire resistance (if not an 'unprotected area') depending on their relationship to the relevant boundary.
Bays or oriels create play platform for children	Where a platform is created within a bay or oriel, care will be required to limit the possibility of a child falling out or through the bay or oriel window. This should be treated as a low-level glazing situation and a means should be provided to prevent any child opening the window.

HAPM GUIDANCE

These features often include a cantilever design which is frequently inadequate. The Technical Audit Unit will expect to see clear drawings and supporting calculations where design is not totally 'robust by inspection'.

Details

COMMON DEFECTS	CRITERIA/GUIDANCE
Gallows brackets and their fixings not durable or not secure under loads	Timber gallows brackets are in contact with the outer leaf and therefore should be treated with preservative and back-painted. The fixings to the outer leaf should be one of: • hot-dip galvanized • austentic stainless steel • non-ferrous.
Support system does not accommodate differential movement between timber frame wall and masonry veneer	Details should show a means of support and soft joints where differential movement is likely between the timber frame wall and the masonry veneer.
No protection for timber columns from water at ground level	Details should be designed so that the end grain is kept dry and can be painted or coated with a preservative.
Rain penetration	Appendages constructed from non-standard joinery and with non-standard details may not be reliably weathertight. For clear guidance on detailing, refer to *Principles of Simple Weathersealed Timber Window Design* published by the Swedish Finnish Timber Council. Note that the exposure of projecting elements is greater than that for a window in the plane of a wall.
Rain penetration at junction of roof with external wall	Tiled roofs to projections should be detailed as any other pitched roof abutment of roof to wall. However, since the size of the roof makes using large interlocking tiles difficult, it is better to use a plain tile. Scaled down ridge and hip tiles are often appropriate where the roof dimensions are small. Flat roofs to projections need to be detailed as any other flat roof or wall abutment (see Topic 4.2 for further guidance).
Cracks at vertical junction with external walls are commonplace	On clay soil, to avoid cracks at the vertical junction with external walls, the projection should have a foundation to the same depth as the main building, or a common raft should be used to avoid the problem. Longitudinal movements can disrupt re-entrant corners at bays in long facades. Refer to BRE *Defect Action Sheet 18* for details of the joint frequency required.
Felt dressed through right angles	Angle fillets should always be detailed to avoid changes in direction in the felt of 90°.
Leaks round glazing	Failure of beaded glazing is very common due to poor glazing practice. Clear guidance is given in *The Glass and Glazing Federation Handbook*.
Cold bridging	Cold bridging at reveals, or where concrete cantilever units are built in, may be a problem with 'add-on' constructions. Therefore, careful consideration of detailing is necessary.

HAPM GUIDANCE

Explicit assembly and arrangement details for these appendages to the main structure should be provided. Basic detailing principles should be applied.

Materials

COMMON DEFECTS	CRITERIA/GUIDANCE
Premature rot in joinery	Timber should be treated with preservative (and back-painted where in contact with the outer leaf).
Pitch of roof too shallow for tiles or slates	As for main roofs, the design requirements in manufacturers' guidance should be used.
Corrosion or expansion of fixing bolts causes cracking	As a minimum requirement, where bolts are to be painted and sealed, they should be galvanized or cadmium-plated, otherwise stainless steel or brass should be used.
Precast concrete components too thin to ensure sufficient cover to reinforcement	When ordinary quality concrete is used, the reinforcement steel normally requires a minimum cover of 30mm, and factory-made units 25mm. Cover is often reduced inadvertently where drips or other indentations are moulded in the unit.
Failure of glazing compound due to drying out	General references to 'non-hardening glazing compounds' often result in materials being used that do eventually harden. All compounds should be referred to by name or with reference to a British Standard, having checked its suitability for the situation and conditions.

5.3

HAPM GUIDANCE

If the appendage is a standard assembly, the manufacturer's quality assurance will be looked for. If, as is more likely, the appendage is an assembly of special items, the Technical Audit Unit will assess its durability in terms of 'life to replacement' and this will be recorded in the audit document.

Workmanship

COMMON DEFECTS	CRITERIA/GUIDANCE
Insecure fixing bolts	Structural fixings should be secure enough to achieve the required design loads. Test loading should be considered. Expanding fixings situated close to edges often fracture the brick or block.
Fixing bolts become loose due to timber shrinkage	Critical connections should be retightened after 6 weeks, providing the timber is dry at that time.
Inadequate fixing of felt at 'outboard' edges	Felt underlays should be fixed along their perimeters by nailing. Aluminium trims should not be used for edges over which water is shed. A felt welted drip should be used.
Inadequate glazing compounds used	Glazing Contractors will often ignore the specification or will formally offer a substitute that is not an 'equivalent' since better quality materials are unpopular, usually being more expensive and difficult to use. Therefore, careful supervision may be required.
Lead flashings ill conceived	Precise leadwork will often not be fully detailed leaving the tradesmen to resolve very tricky arrangements. In the absence of designed details, the guidance in the *Lead Development Handbook* should be followed.
Mastic joints poorly formed	Care should be taken to avoid the following common problems: • incorrect width-to-depth ratio • no bond break tape • no joint backing strip • poor tooling.

HAPM GUIDANCE

The Technical Audit Unit will pay special attention to workmanship associated with building and attaching appendages as these are often poorly considered and often result in defects.

Design

COMMON DEFECTS	CRITERIA/GUIDANCE
Stairs too steep	*Approved Document K* specifies that the angle of pitch should not exceed: • 42° for a private stair (ie serving only one dwelling) • 38° for common stairs. *Approved Document K* also gives the requirements for the maximum rise and minimum going. Where the angle of pitch is less than 15°, a ramp can be used.
Insufficient headroom to stairs	*Approved Document K* specifies a minimum headroom of at least 2.0m above the pitch line.
Insufficient width to stairs	*Approved Document K* specifies the minimum unobstructed widths for private stairs, ranging between 600mm and 900mm.
Obstruction of stairs or landings by doorways	Doorways should not open directly onto stairs. Landings should be wide enough to accommodate a 400mm wide space clear of any door swing at the top or bottom of all flights. Risers in any one flight are limited to 16, so buildings with high storey heights may require landings in mid-flight. BS 5395: Part 1: 1977 *Stairs, ladders and walkways: Code of practice for the design of straight stairs* gives guidance on suitable layouts for stairways.
No handrails provided	All flights and ramps should have handrails on at least one side, the height of which is specified according to the type of stair or ramp used. For institutions catering for young children and, if possible, for family housing, additional handrails should be provided at a lower level (about 600mm above the pitch line). *Approved Document K* and BS 6399: Part 1: 1984 *Loading for buildings: Code of practice for dead and imposed loads* specify the design loadings required for handrails and balustrades. BS 6180: 1982 *Code of practice for protective barriers in and about buildings* recommends guarding a drop of 500mm. However, *Approved Document K* gives a mandatory requirement for guarding a drop of 600mm.
Inadequate lighting of public access balconies, stairs and ramps	Lighting should be sufficient to highlight the top and bottom of all flights and ramps and to eliminate all dark corners in localities where mugging may be prevalent. Provision for emergency lighting is required in accordance with BS 5266: Part 1: 1988 *Emergency lighting: Code of practice for the emergency lighting of premises other than cinemas and certain other specified premises used for entertainment.* Refer also to Section 6.1 in CP 3: Chapter IV: Part 1: 1988 *Code of basic data for the design of buildings: Precautions against fire: Flats and maisonettes (in blocks over two storeys)* for the requirements for stairway and corridor lighting.
Provision of single steps	Single steps can cause accidents and therefore should never be used.
Surface ponding of water	BS 6367: 1983 *Code of practice for drainage of roofs and paved areas* gives guidance on the design of drainage for paved areas. Sufficient falls and outlets should be provided.

Design

COMMON DEFECTS	CRITERIA/GUIDANCE
No movement joints in balconies	All movement joints should be shown on the construction drawings. Consideration should be given to the practical problems of forming and inspecting such joints. The spacing of movement joints depends on the type of material used in their construction but, generally, a movement joint at each alternate dwelling in long runs of blocks of flats will be sufficient.
Balustrades too flimsy	Balustrades on common stairs and balconies should be robust. BS 6399: Part 1: 1984 specifies the design loading for each class of use.
External stairs waterproofed with asphalt	Stairs may need to be waterproofed where they also form a roof over an internal space. Experience has shown waterproofing can fail where exposure to ultra-violet light is significant. Since non-slip nosings are difficult to incorporate and the asphalt often slumps from the vertical risers, it is best to locate the waterproof membrane under the stair flight.

HAPM GUIDANCE

All stairs and ramps should conform to the requirements of *Approved Document K.* Refer to BS 5395: Part 1: 1977, BRE *Defect Action Sheet 53* and to CP 3: Chapter IV: Part 1: 1988.

Details

COMMON DEFECTS	CRITERIA/GUIDANCE
Inadequate fit of components	Stair flights, landings and walkways are often prefabricated. The joints between the prefabricated components and supporting structures should be detailed to allow for the maximum combination of manufacturing and erection tolerances.
Inadequate 'freeboard' at upstands and under door thresholds	Waterproofing should be detailed to prevent water penetration into dwellings when balcony outlets become blocked. Particular care is required where the deck level of the balcony is above the floor level of an adjoining dwelling.
Buckling of long runs of steel handrail	Movement joints should be provided at convenient locations on the steel handrail to suit the design of the supporting structure.
Cracking in brickwork or blockwork balustrades	Thermal movement can cause cracking to occur in brickwork or blockwork, therefore movement joints should be provided to match those in the supporting structure. In long runs of balustrade, movement joints should be provided at maximum intervals of about 6 metres.
Waterproofing perforated by balustrade supports	Balustrade supports fixed into or through the waterproofing should be detailed to enable the Builder to achieve a practicable watertight arrangement. Asphalt should be dressed up around sockets, and skirts to shed water should be provided on standards. Note that asphalt is liable to crack under any suddenly-applied strain, but particularly under impact in cold weather.
Asphalt not isolated from substrate movement	Asphalt should be separated from its substrate by an isolating membrane of sheathing felt to bridge discontinuities. As a precaution, expanded metal should be added over potential stress areas, such as the joints between different materials. A formal joint should be provided where significant movement may occur.
Water staining on face of building	To avoid unsightly water staining, drips and outlets should be detailed so that water is shed clear of the faces of the balustrade and of the building.
Cracking and curling of surfaces at junctions of different materials	Since different materials have different rates of expansion and contraction, allowance should be made at any joint to avoid cracking.
Movement joints not carried through surface finishes	Any movement joint in the supporting structure should be continued through the finishes and surfacing to avoid early failure of the finishes.
Dpc bridged by ramped external surfaces	Ramps to doors should be set away from adjacent walls by at least 75mm.

HAPM GUIDANCE

As with porches and bay windows, external stairs, balconies and ramps should be detailed carefully to avoid defects which can result in danger to users.

Materials

COMMON DEFECTS	CRITERIA/GUIDANCE
Stairs and walkways too slippery	Many materials are available for the surfacing of external balconies, stairs and ramps but few offer sufficient slip resistance when wet. The following materials are the most suitable: • clay tiles with carborundum finish • clay tiles with textured finish • mastic asphalt with dimpled finish • concrete with textured finish • granolithic with carborundum finish. Non-slip nosings should be provided on stairs.
Brick or stone copings on balustrades subject to frost attack and movement	Unless great care is taken in specifying the mortar for setting copings and in the actual workmanship to ensure that all joints are completely filled, there is the possibility of moisture penetration into the joints leading to the movement and dislodgement of the copings during frost. There is a similar risk of movement of the copings under cyclic thermal action if inadequate provision has been made for movement joints.
Inadequate protection of steel components	Steel is widely used for railings and balustrades and has a very limited life if not protected sufficiently and maintained. All steel used in exposed locations, such as railings and balustrades, should be hot-dip galvanized after fabrication to conform with the requirements of BS 729: 1971 *Specification for hot dip galvanized coatings on iron and steel articles* (to leave a zinc deposition of 610g/m² before a paint treatment is applied). Newly-galvanized steel should be etch-primed or coated with mordant solution before painting. Consideration should be given to the access required to any areas for maintenance.
Inadequate cover to reinforcement of concrete items	Concrete members in stairs and balustrades are usually of a small cross-section making the minimum cover to reinforcement difficult to achieve in practice. Where the thickness of members cannot be increased, additional anti-corrosion protection should be provided to the reinforcement.
Inadequate resistance of adhesives or beddings to frost	Manufacturers should always confirm that their products are suitable for the proposed conditions.
Light fittings subject to vandalism	Light fittings for use in public places should be vandalproof. Fixings and means of access to the fittings should not be visible.
Handrails subject to vandalism	All fixings should be made vandalproof and, as far as possible, should be located out of sight.

Materials

COMMON DEFECTS	CRITERIA/GUIDANCE
Asphalt surface lacks surface protection	Plain or dimpled asphalt may not be sufficiently durable in areas of heavy usage or where point loads occur. The provision of promenade tiles or anti-slip clay tiles should be considered, or the addition of extra grit to roofing grade asphalt.

HAPM GUIDANCE

Common stairs, ramps and balconies have to withstand considerable wear and tear and abuse – often in exposed conditions. Historically, external components in such locations have performed poorly in terms of durability. It is not necessarily sufficient that quality of materials and standard of workmanship should comply with the relevant British Standards or be supported by a BBA Agrément Certificate. Materials should be of exterior quality and vandalproof.

Workmanship

COMMON DEFECTS	CRITERIA/GUIDANCE
Errors in setting out	All the risers in any stair flight should be of equal height. In the case of concrete construction, when there may be differing thicknesses of finishes on floors and treads, particular care should be exercised in setting levels.
Inadequate attention to stability in temporary conditions	To avoid conditions of temporary instability, particular care is necessary, especially when prefabricated stair flights are being installed. Narrow temporary bearings in loose-fitting support frames with inadequate structural ties can present an unacceptably high risk. Therefore, the method for erecting the stairs should be fully considered beforehand.
Fixing bolts insecure	Loose connections can lead to the premature failure of connections, and badly finished fixings can lead to unsightly staining of the structure. All fixing bolts should be fully tightened, and sockets filled and grouted. Exposed nuts should be spot-welded to prevent loosening by vandals.
Structure overloaded by temporary storage of materials	Constant care should be exercised to ensure that balconies and walkways are not overloaded by storing materials on them during construction. Note that balustrades, and their fixings, can easily be damaged during construction by crane loads being allowed to swing against them.
Inadequate protection of new concrete	Newly-poured concrete should be protected: • in hot weather against too rapid drying out to prevent shrinkage cracks • in cold weather against frost damage or freezing. The method of curing the concrete should always be given careful attention.
Inadequate protection of new brickwork or blockwork	As with concrete, new brickwork and blockwork should be protected from: • inclement weather to prevent the mortar being washed out • frost damage.
Uneven surface finish	Any unevenness of stair or walkway finishes is a hazard. Inadequate attention to accurate levelling can result in ponding of water in wet weather leading to water penetration into dwellings, or to the formation of ice in cold weather.
Inadequate attention to laying of finishes	The preparation of the surface on which finishes are to be laid should be carried out with care if the surfacing material is to be sufficiently durable. In the case of proprietary materials, the manufacturer's instructions should be followed strictly.
Inaccurate placing of top reinforcement in cantilevers	There have been a number of serious structural failures due to top reinforcement being misplaced or supported inadequately during concreting.

HAPM GUIDANCE

Dimensional accuracy is of prime importance, followed by extensive site supervision to ensure a high standard of workmanship. The Technical Audit Unit Site Inspectors will undertake a 'quality count'.

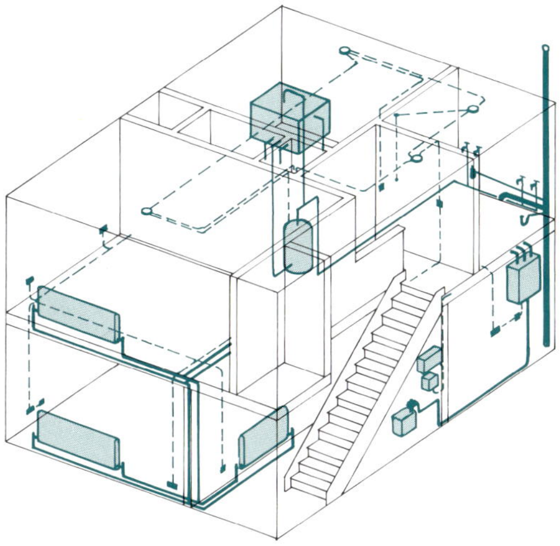

ELECTRICAL INSTALLATION

HOT AND COLD WATER INSTALLATIONS

GAS HEATING

ABOVE GROUND DRAINAGE

SHAFTS, DUCTS, CHUTES

6

HAPM

HOUSING ASSOCIATION PROPERTY MUTUAL LIMITED

Design

COMMON DEFECTS	CRITERIA/GUIDANCE
Inadequate socket provision and location	A sufficient number of socket outlets should be specified and sensibly located to avoid using adaptors and long trailing flexes being used by occupants. Double sockets should be provided in preference to single sockets. For general-purpose housing, the recommendations in *Homes for the future* published by the Institute of Housing and RIBA represent an adequate standard.
Inadequate provision for task lighting	The location of lighting points is important in living rooms and kitchens where, generally, the normal single central point is inadequate. In living rooms, several lighting points should be located in corners or over the likely position of the dining table. In kitchens, several lighting points should be located strategically over worktops.
Inadequate lighting of stairs and landings	Refer to BS 8206: Part 1: 1985 *Lighting for buildings: Code of practice for artificial lighting* which recommends good lighting levels to highlight treads, risers, landings and half-landings using luminaries selected and positioned to avoid direct glare to users.
Electrical intake and meter positions do not conform with supply company requirements	Supply companies have enforceable requirements concerning the position and accessibility of their service head and main fuse, and also their meters. These should be checked before the design is completed.
Inadequate provision for earthing	In modern construction, the water main is normally polyethylene and therefore it cannot be used as an earthing point. Three methods of earthing are commonly used: • via the sheath of the intake cable (TNS) • via the neutral conductor of the intake cable (TNC-S) • by means of an earth conductor (TT) driven into the ground. Note that such electrode pins will require periodic checking to confirm their efficiency which can vary with the extent of ground moisture.
Inadequate provision for equipment bonding	The fifteenth edition of the *IEE Regulations* requires the equipotential bonding of: • 'exposed conductive parts' (parts of the electrical installation liable to become live under fault condition) • 'extraneous conductive parts' (metal parts not part of the electrical installation, such as radiators, metal windows, baths, etc). Explanations of these bonding requirements are contained in various handbooks explaining the *IEE Regulations*.
No de-rating of cables covered by insulation	Cables covered by insulation can be subject to overheating and should be de-rated by a factor of 0.5 (insulation on both sides) or 0.75 (insulation on one side). Refer to BRE *Defect Action Sheet 62* for details.
Inadequate protection against electric shock	The *IEE Wiring Regulations* now require residual current operated circuit breakers (RCCBs) for protecting all 'TT' ring main circuits, and for any socket supplying appliances outside the equipotential zone. Where RCCBs are used, special circuits may be required to protect against disconnection of the fridge or freezer.

Design

COMMON DEFECTS	CRITERIA/GUIDANCE
Incorrect sizing of electric storage heaters	Electric off-peak storage heaters will give acceptable heating costs only if they are installed in well-insulated dwellings. The insulation should be at least in line with the current standards given in the *Building Regulations* and should also include draughtstripping of the doors and windows. Where a heating consultant is not employed, the Medallion Award Scheme sponsored by the local electricity company can provide heat loss calculations and work out a suitable heating scheme. To avoid running out of heat in the early evening, sizing of the heating system should err on the generous side. Note that modern electric combi-heaters incorporate a thermostatically-controlled on-peak element which can provide an automatic top-up when required. The minimum size of storage heater generally available was about 1.7kW, which is too large for smaller rooms. Therefore, it was usual to supplement off-peak heaters with on-peak panel heaters or downflow heaters in bathrooms, kitchens or second and third bedrooms but this leads to excessive running costs. However, now at least one manufacturer makes a 900W storage heater suitable for bathrooms and small bedrooms.
Incorrect siting of electric storage heaters	Although electric storage heaters are getting more compact, care should be taken to locate them where they will cause the least inconvenience. Storage heaters should be located on internal walls to minimise the heat loss. Storage heaters should not be located at the mid-span position of floor joists and beams because of their weight, unless provision has been made for the resulting dead load. Where electric storage heaters are designed to heat two spaces, for example a living room with connected kitchen, they should be placed centrally in the combined space.
Insufficient control systems for electric storage heaters	Control of electric storage heating consists of regulating the length of time the unit receives a charge and/or regulating the rate at which heat is lost from the heater during the day. Controls are getting much more sophisticated and duration-of-charge controllers can be linked to outside temperature sensors. The rate that heat is released from the storage heaters is controlled by flaps, dampers and fans which can also be controlled by remote thermostats. *DOM 11* published by the Electricity Council explains the theory behind storage heater controls.

HAPM GUIDANCE

Correct installation of electrical works has implications on the health and safety of future users of the building. The importance of a safe electrical installation complying with the appropriate regulations cannot be overemphasised. The comments on lighting and socket outlet provisions in this topic are for guidance only. However, underprovision of such facilities, in the interests of economy, may encourage informal adaption and extension of the system by occupants. All electrical installations should be covered by a National Inspection Council for Electrical Installation Contractors (NICEIC) Completion Certificate.

Designers of electric storage heating systems should ensure that the system is designed competently by either the electricity authority or a heating consultant. The internal design temperatures should be specified and an assurance should be sought that these temperatures can be maintained for the whole day under normal conditions of use without recourse to on-peak electricity. When specifying electric storage heating, every effort should be made to increase thermal insulation levels as much as possible over the minimum requirements of the *Building Regulations*.

Details

COMMON DEFECTS	CRITERIA/GUIDANCE
PVC cables in contact with expanded polystyrene insulation	Prolonged contact between PVC and expanded polystyrene will result in plasticiser migration from the PVC leading to embrittlement of the sheathing and an increased fire risk.
Inadequate provision for cable protection routing and rewiring	A fully rewirable system in metal conduit is rarely specified for domestic work, therefore some destructive work will be necessary when a rewire is required. Normally, metal conduit is used for casting into concrete floor slabs for electrical services but a better solution is to use sheathed cable in plastics conduit located in a dropped ceiling. In accessible loft spaces and timber suspended floors, sheathed cable clipped to the sides of joists and rafters is acceptable. In walls behind plaster or dry lining, sheathed cables should be protected by a metal channel and run vertically in all cases.
Socket outlets awkward to use	Outlets should be located at least 150mm above the floor level (1000mm for buildings used by the elderly and disabled). In kitchens, some outlets should be located below worktop level with switches above the worktop. Extra sockets should be specified to take account of permanently connected appliances.
Consumer unit out of reach	It is very common to locate the consumer unit in a position where it is very difficult to reach, such as above a door or similar. Note that while the consumer unit should not be readily accessible by children, occupants should be able to change a fuse cartridge or re-set a trip in the dark without needing to climb on chairs or ladders. The recommended height for mounting the consumer unit is 1.5 metres.
Light pendants clash with door swings	The problem of light pendants clashing with door swings should be checked when detailing. A bulkhead fitting can solve the problem.
Non-fused spurs overloading cable	The installation of non-fused spurs should be avoided unless it can be determined that there is no likelihood of the spur cable being overloaded.

HAPM GUIDANCE

Normally, the positions of electrical equipment are shown diagrammatically on a plan layout. Special details may be appropriate to show arrangements in kitchens (ie below and above the worktop), positioning of the consumer unit and special cable routing requirements.

Materials

COMMON DEFECTS	CRITERIA/GUIDANCE
Incorrect sizing of earthing and bonding cables	All cables should be sized correctly in accordance with the current edition of the *IEE Regulations*. Depending on whether or not mechanical protection is provided, the minimum cross-sectional area of bonding cables should be one of: • 2.5mm^2 • 4.0mm^2 • 6mm^2.
Incorrect sizing of circuit cables	Generally, the following sizes of circuit cables should be used: • 2.5mm^2 for ring main circuits • 1.5mm^2 for lighting circuits • 4mm^2 or 6mm^2 for cooker circuits, depending on the cooker loading • 2.5mm^2 for immersion heaters up to 3kW.
Equipment non-standard or not complying with IEE Regulations	Only materials complying with the relevant British Standards should be used for electrical installations. Beware of sub-standard imports and substitutions during the work.
Final connections to heat-producing appliances not in heat-resisting cable	The final connections to heat-producing appliances should be specified correctly and checked on site.

HAPM GUIDANCE

All components, materials and assemblies should comply with the appropriate British Standards and the *IEE Regulations*. Any such materials selected should ensure satisfactory service for the life of the system. Under normal circumstances, it is envisaged that properly installed, PVC-sheathed cables will last for the period of the Major Defects Insurance Policy. Fittings will not be so durable and will require replacement periodically, certainly once in the life of the Policy.

Workmanship

COMMON DEFECTS	CRITERIA/GUIDANCE
Inadequate support to cable	Refer to Appendix II in the *IEE Regulations* which details the methods of supporting various types of cable, with appropriate centres.
Insulation stripped carelessly	The outer sheathing of any cable should be visible inside the mounting boxes, and no core insulation should be visible outside the fitting.
Accessories not level	To accommodate accurate levelling, most modern fittings have elongated mounting holes.
Dirty or painted accessories	Fittings should be removed prior to decoration and refixed afterwards. This calls for some co-ordination of trades, or a postponement of the electrical second fix until after decoration. Note that there is no reason why the decorators should not remove and replace fittings.
Poorly installed mini-trunking	The appropriate mini-trunking fitting should be used in all situations, such as internal and external corners and connections to fittings.
Inadequate bonding connection	Special bonding connectors should be used. Any coating or paint on the pipe to be bonded should be scraped off to achieve electrical continuity.
Not enough cable slack left at fittings	Sufficient slack should be left in the cable to enable the fitting to be easily connected and renewed.
Knock-out holes in metal flush boxes not protected with rubber grommets	To avoid the risk of cables chafing, all knock-out holes in metal casings and flush boxes should be protected with rubber grommets.
Metal flush boxes not earthed	The socket outlet should be connected to the metal flush box with a sleeved earth wire to provide continuous protection, even when the fitting is removed.
Earthing and bonding cables not labelled	All earthing and bonding cables should be labelled close to the point of connection with "SAFETY ELECTRICAL CONNECTION – DO NOT REMOVE".
Cables accommodated by notching the top of joists	Cables should be mounted through holes drilled through the centre of the floor joists. Where this is not possible, and notching the top of the joists is the only alternative: • the cables should be protected from damage by a metal plate thick enough to resist nail penetration • the notches should be the minimum size possible and located in the end third of the joists.
Circuits not identified at consumer units	At the consumer unit, each fuse, MCB and RCCB should be clearly identified.

HAPM GUIDANCE

Installers and Supervisors should pay particular attention to these and other similar common problems. The Technical Audit Unit Site Inspectors will undertake a 'quality count'. Generally, the Electrical Contracting industry operates a dependable quality assurance system and an 'Inspection Certificate' is issued for each installation. Copies of these documents should be provided for retention by Supervisors.

Design

COMMON DEFECTS	CRITERIA/GUIDANCE
Low mains pressure	The water companies have a duty to provide water at a minimum pressure of their own choosing. This pressure may be insufficient to provide an acceptable flow at taps. The local water supply company should be contacted to find out their minimum declared mains pressure and to ensure it is sufficient. Where the mains pressure is insufficient, consideration should be given to boosting the water pressure with a break tank and pump. In low water pressure areas, the use of mains-fed combi-boilers should be avoided.
Low water pressure resulting from limited static head	Low water pressure may be a particular problem in flats where the cold water tank is accommodated on the same storey. Low water pressure can limit the effectiveness of showers and can cause problems with automatic washing machines. The problems of low water pressure can be avoided by: ● fitting an in-line pump operated by a flow switch ● using a mains-fed secondary hot water system ● using a combi-boiler. The problems of low water pressure on showers can also be solved by using a proprietary cylinder which incorporates a mains-fed heat exchanger connected to the shower circuit. Using long pipe runs or pipes of insufficient size can contribute to the problem of low water pressure.
Incorrect sizing of hot water tank	Continuously heated hot water storage cylinders, as with gas-fired central heating systems or on-peak electric systems should have a capacity of at least 150 litres for a 3 bedroom house. Reduced storage capacity may be considered for smaller dwellings. Where the water heating system depends on intermittent heating, such as off-peak electric systems, the capacity of the hot water system should be increased to 210 litres for a 3 bedroom house and 140 litres for a 1 bedroom house.
Excessively long 'dead-leg' pipework to taps	To minimise the waste of water BS 6700: 1987 *Specification for design, installation, testing and maintenance of services supplying water for domestic use within buildings and their curtilages* makes recommendations which limit the length of uninsulated 'dead-legs' to 12m for 15mm and 22mm diameter pipes. In excess of these lengths, the pipes should be insulated. However, 'dead-legs' are best avoided and should be insulated whenever practical.
Poor siting of tanks and insufficient access for maintenance	Where hot and cold water tanks are to be sited in cupboards, sufficient access space should be provided around the tank to enable the fitter to make the connections and for maintenance. There should be a space of at least 300mm above the tank to enable the ball valve to be serviced or replaced. In small cupboards, it may be necessary to provide a removable panel above the door head to allow access to the tanks for maintenance, as recommended by BS 6700: 1987.

Design

COMMON DEFECTS	CRITERIA/GUIDANCE
Inadequate protection against mains water contamination	With the increasing use of mains-supplied appliances now allowed by the *Water Supply Byelaws*, suitable precautions should be taken to prevent contamination of the mains water. This can be achieved by using: • an air gap • a check valve • another device for preventing back flow. *Water Supply Byelaw 11* describes acceptable devices to prevent mains water contamination.
Cold water in storage tank gets warm	The need to avoid overheating of stored cold water is not always recognised. The storage temperature of cold water should be below 20°C (except in feed and expansion tanks for heating). The cold water tank should be located in a cool environment, taking account of any likely solar gain.

HAPM GUIDANCE

Designers should ensure that the hot and cold water installation complies with the requirements of the water supply company.

Details

COMMON DEFECTS	CRITERIA/GUIDANCE
Hard water deposits accumulate in mains-fed boilers and cylinders	In hard water areas, carbonate or sulphate accumulation may be a problem causing 'furring up' of pipes and reducing the efficiency of heat exchangers. To allow the periodic introduction of chemicals to dissolve the deposits that form in water installation in hard water areas, the installation of T-fittings and valves should be considered.
Inadequate provision for overflow or warning pipes	Overflow pipes should: • be supported adequately (this can be difficult in roof spaces) • be insulated • have a continuous fall away from the tank • discharge in a conspicuous position. *Water Supply Byelaw 30(2)* requires overflow pipes to be screened against the entry of insects. Special overflow fittings incorporating mesh filters are available. Note that inside the tank, the overflow pipe should be turned down below the level of the water to limit the entry of cold air.
Installation susceptible to freezing	To avoid freezing of the water supply, external underground mains water pipework should be laid at a depth greater than 750mm. Where possible, pipework should not be located in unheated parts of the dwelling or, where this is not possible, the pipes should be insulated. In extreme cases, trace heating controlled by a frost thermostat can be provided. Tanks in roof spaces should be insulated all round, but not underneath in order to benefit from heat loss through the ceiling. Full guidance on frost protection is given in Clause 10 in BS 6700: 1987 *Specification for design, installation, testing and maintenance of services supplying water for domestic use within buildings and their curtilages*.
Cold water storage tanks supported inadequately	Plastics and glass fibre tanks should be provided with a continuous support platform of timber boarding or plywood. Rigid metal tanks can be supported on bearers placed at 300mm centres. In trussed rafter roofs, the weight of the tank should be spread and transmitted to the rafters as close as possible to the node points as explained in the International Truss Plate Association's *Technical Bulletin No 4*.
Inadequate provision of stopvalves, service valves and draincocks	*Water Supply Byelaws 61, 62* and *63* govern the provision of stopvalves on the incoming water main. The general requirements are that the stopvalves should be easily accessible. *Water Supply Byelaws 65, 66* and *67* require a drain tap to be provided immediately above the stopvalve to enable easy draining of the water in the service pipe. *Water Supply Byelaws 68, 69* and *70* govern the provision and location of service valves controlling the supply of water to and from: • appliances • hot and cold water tanks • float-operated valves.

Details

COMMON DEFECTS	CRITERIA/GUIDANCE
Water main pipework conducive to condensation	The cold surface of water mains pipework can result in condensation problems, especially when the water service enters the dwelling in an unventilated location, such as an undersink cupboard. Pipework should be insulated using closed-cell, preformed pipe insulation.

HAPM GUIDANCE

Designers are advised that non-compliance with the requirements of the *Water Supply Byelaws* can, in certain instances, lead to prosecutions. Designers should provide a schematic layout of the hot and cold services indicating the location of all valves, fittings and tanks, etc. In the absence of precise details, one installation should be completed and 'approved' on site as the model for all others.

Materials

COMMON DEFECTS	CRITERIA/GUIDANCE
Use of solders containing lead	*Water Supply Byelaw 7* prohibits the use of materials likely to cause contamination in pipework carrying potable water. In particular, lead-free solders should be used in all capillary fittings. However, although these solders are now available on the market, they are not yet covered by a British Standard so vigilance is required on site to identify the type of solder being used.
Omission of lids to tanks	*Water Supply Byelaw 30(2)* requires cisterns storing water for domestic purposes to have a lid which excludes light and insects but is not airtight. Any lid should be cut to fit closely around vents or expansion pipes discharging into the tank. Note that proprietary purpose-made lids are now available providing screening of air inlets and provision for sealing around the pipes.
Corrosion of galvanized tanks	Plastics or glass fibre tanks should be used in preference to galvanized steel tanks.
Corrosion of pipework due to electrolytic action between dissimilar metals	Generally, copper pipe should not be used upstream of lead pipe. In addition to the nature of the dissimilar metals, the quality of the water also has an effect on the susceptibility of the pipework to corrosion. Refer to BRE *Digest 98* and BS 6700: 1987 *Specification for design, installation, testing and maintenance of services supplying water for domestic use within buildings and their curtilages* for guidance.
Use of non-BS-kitemarked copper pipe	BS-kitemarked tube is the only copper tube which complies with *Water Supply Byelaw 52*. To avoid problems with carbon film deposits causing pinhole corrosion in certain waters, copper tube should be specified to Table X or Table Z in BS 2871: Part 1: 1971 *Specification for copper and copper alloys. Tubes: Copper tubes for water, gas and sanitation.* Tube to Table Z (thin wall) should never be bent. For installations in the ground, tube to Table Y should be used.
Corrosion of copper cylinders	To prevent corrosion in areas with aggressive mains water, copper cylinders should be fitted with a sacrificial anode of aluminium alloy.
Plastics pipe clips allow pipe to 'pop out'	Clips with a band across the front of the pipe hold the pipe more securely than push-fit clips. Generally, 15mm copper pipework should be clipped at 1200mm centres horizontally and 1800mm centres vertically.
Inadequate resistance of brass fittings to dezincification	To comply with *Water Supply Byelaw 53*, all brass fittings used underground, or in an otherwise inaccessible location, should be non-dezincifiable. Where this is not possible, gun metal fittings should be used. In other locations, *Water Supply Byelaw 52* requires the use of dezincifiable-resistant fittings only in certain water areas. To avoid problems of unauthorised substitution, consideration should be given to specifying non-dezincifiable fittings throughout. Note that non-dezincifiable fittings may be difficult to obtain readily.

Materials

COMMON DEFECTS	CRITERIA/GUIDANCE
Incorrect specification for cylinder grade	BS 1566: Part 1: 1984 *Copper indirect cylinders for domestic purposes: Specification for double feed indirect cylinders* describes three grades of cylinder (Grade 2, Grade 3 and Grade 4: • Grade 2 cylinders are suitable for most domestic installations which have a cold water head exceeding 10m but less than 15m • Grade 3 cylinders are suitable for most domestic installations which have a cold water head of less than 10m. Although Grade 4 cylinders are available, their use should be avoided.

HAPM GUIDANCE

Designers should ensure that specifications cover adequately the above items which may be difficult to check on site.

Workmanship

COMMON DEFECTS	CRITERIA/GUIDANCE
Pipework not cleaned before use	*Water Supply Byelaw 55* requires that every pipe which carries water for domestic use should be flushed out before it is used to remove debris, such as flux residue or filings which can cause float valves to stick open.
Pipework fails hydraulic test	Underground or inaccessible pipework should be capable of withstanding at least twice the normal working pressure. All other pipework and fittings should be capable of withstanding one-and-a-half times the normal working pressure. All completed installations should be subjected to a hydraulic test.
Drinking water supply warmed by adjacent hot pipe	Hot and cold water pipes should be separated sufficiently or, where this is not possible, the pipes should be insulated.
Pipework joints not made well	Pipe ends should be cut square and pushed fully home into fittings. Internal burrs should be removed and all excess flux and solder wiped off.
Blockages in pipes	During construction, all pipe ends should be taped over to prevent foreign bodies entering the pipe accidentally (or wilfully).

HAPM GUIDANCE

As designers are often reluctant to fully detail domestic service installations, more responsibility rests with Contract Supervisors to ensure the adequacy of installations. The requirements of the new *Model Water Byelaws* are far reaching and therefore it would be prudent to invite a representative from the local water company to inspect a sample installation.

Design

COMMON DEFECTS	CRITERIA/GUIDANCE
Incorrect sizing of boiler and system	The size required for the gas heating system and the boiler is determined by reference to: • minimum expected external temperatures • desired internal temperatures • level of heat loss • degree of intermittency • hot water requirements. Some problems can arise with very small and/or highly insulated dwellings because radiators with very low outputs are not available. General guidance on sizing the system is contained in the British Gas publication *A guide to the design and operation of domestic wet central heating systems.*
No allowance for unheated attached dwellings	Generally, it is left to specifiers to decide whether to design the system to cope with unheated attached dwellings. Many specifiers think it is prudent to make some allowance, such as providing for a 10°C temperature differential through all adjoining floors and walls.
Incorrect siting of flue terminal	Guidance on the siting of balanced and open flues in relation to windows, ventilators and obstructions is contained in the British Gas publication *A guide to the design and operation of domestic wet central heating systems.* The range of acceptable flue locations can be increased by using a fan-assisted flue.
Poor siting of boiler	Boilers should be fully accessible for maintenance. The casings of some boilers may become too hot to be placed close to kitchen wall units or fridges (the manufacturer's installation instructions should be checked). Open-flued boilers should not be sited in: • bathrooms • bedrooms • bedsitting rooms.
Inadequate provision for ventilation	Open-flued boilers require a permanent supply of air for combustion. Compartments containing room-sealed boilers or appliances also require ventilation. Guidance on ventilation requirements is contained in the British Gas publication *A guide to the design and operation of domestic wet central heating systems* and also in BS 5440: Part 1: 1990 *Installation of flues and ventilation for gas appliances of rated input not exceeding 60kW (1st, 2nd and 3rd family gases): Specification for installation of flues.*
Poor siting of radiators	Where possible, to overcome cold down draughts, radiators should be sited under windows. This is less important with double-glazing and the under window location can lead to problems with: • long curtains • constraints on radiator sizing • increased heat loss. Generally, the higher the level of thermal insulation the less important the radiator siting becomes.

Design

COMMON DEFECTS	CRITERIA/GUIDANCE
Insufficient static head for boiler	The static head requirement is generally specified by the boiler manufacturer. The static head requirement can be difficult to achieve with wall-mounted boilers in flats. In cases of difficulty, consideration should be given to using a sealed, pressurised system.
Incorrect siting of pump/ open vent/cold feed	Failure to achieve the correct relationship between the positions of the pump, open vent and cold feed can result in air entrainment, and noise and water being discharged from the open vent. The boiler manufacturer's installation instructions usually give specific advice on the relative positions of the pump, open vent and cold feed, but general guidance is given in the British Gas publication *A guide to the design and operation of domestic wet central heating systems.*

HAPM GUIDANCE

Designers will be required to supply heat loss calculations to confirm the correct sizing of the heating system and to confirm that the system is designed in conformity with the requirements of the manufacturers of the principal components. Designers should heed all advice in the British Gas publication *A Guide to the design and operation of domestic wet central heating systems.*

Details

COMMON DEFECTS	CRITERIA/GUIDANCE
Incorrect siting of room thermostats	Room thermostats should be sited away from the effect of short-term or localised changes in air temperature. Therefore, the following locations should be avoided: • over radiators • in areas subject to solar gain • on external walls • in rooms with supplementary heating.
Unprotected pipes in screeds	Generally, pipework should not be buried in screeds, but where this is unavoidable, eg across doorways, pipework should be protected by a plastics sleeve and should not contain any joints. Note that it would be better for pipework ducts to be included within the screed.
Use of aggressive fluxes	Since flux residues are difficult to remove, corrosive or 'self-cleaning' fluxes should not be used.
Inadequate provision of valves and fittings for maintenance	It should be possible to remove major components, such as pumps and motorised valves, for replacement or maintenance without draining the system, by fitting isolating valves on either side of the component.
No provision for a permanent by-pass	Most low-capacity boilers are required to be fitted with a valved by-pass and a pump over-run device to ensure some circulation after boiler shutdown to prevent 'kettling'. A by-pass is also required in systems where all the radiators are fitted with thermostatic radiator valves.
Non-compliance with specific water company requirements	On all mains-fed and sealed, pressurised systems, to prevent contamination of the mains water, strict compliance with the requirements of the local water company is necessary. The *Water Supply Byelaws Guide* published by the Water Research Council contains much useful guidance.
Inadequate fixing of boilers	Proper fixing of wall-mounted boilers is essential. Note that fixing difficulties can be encountered in timber framed buildings and in buildings with dry lined walls.
Excessive notching of joists for pipework	Notches should not be deeper than one-eighth the depth of the joists and should be confined to the end third of the joists. Refer to BRE *Defect Action Sheet 99* for guidance.
Inadequate provision for venting and draining	All high points in the pipework circuit should be provided with manual or automatic air vents. Low points in the pipework circuit should be fitted with draincocks.

Details

COMMON DEFECTS	CRITERIA/GUIDANCE
No balanced flue terminal guards	The *Building Regulations* require that when a balanced flue terminal could come into contact with persons, or could be subject to damage, it should be protected by a guard.

HAPM GUIDANCE

It is common to leave pipe layout and component positioning to the discretion of the Plumber/Heating Engineer. Consequently, often every house on an estate has a different pipework layout. Pipe layouts should be provided and approved by the person responsible for structural design. Associated details should show the position of the boiler and precise positioning of tanks, pumps and motorised valves. Alternatively, contract documentation should provide for a 'sample' installation to be approved on site in advance of further installations.

Materials

COMMON DEFECTS	CRITERIA/GUIDANCE
Corrosion of radiators	Little is known about the pinhole corrosion of radiators, but research is continuing. The normal internal corrosion of steel radiators proceeds too slowly to cause problems. Note that an excessive build-up of iron oxide in radiators can cause problems with the pump and indicates that air is entering the system.
Corrosion of galvanized steel tanks	Localised corrosion of galvanized tanks can be caused by the electrolytic action set up between steel swarf and the zinc galvanizing. Generally, only plastics or glass fibre water tanks should be used.
Corrosion of pipework	Since some non-British Standard tube is susceptible to pinhole corrosion, copper pipework should be specified to Table X in BS 2871: Part 1: 1971 *Specification for copper and copper alloys. Tubes: Copper tubes for water, gas and sanitation.* The British Standard kitemark should be checked. Corrosion caused by swarf and aggressive fluxes is more common in 'dead-legs' where cleansing by water circulation is not effective. External corrosion of copper pipework can be caused by cement, therefore pipework buried in cement screeds should be protected by a plastics coating, or wrapped in isolating tape.
Corrosion of cylinders	To prevent corrosion in areas with aggressive mains water, copper cylinders should be fitted with an aluminium alloy sacrificial anode.
Leaking radiator valves	Cheap radiator valves have a tendency to leak around the spindle and at the union with the radiators. Only valves designed and manufactured to BS 2767: 1972 *Specification for valves and unions for hot water radiators* should be used.
Noisy systems	Noise from central heating systems can be caused in numerous ways. Some common sources of noise to be avoided are: • poorly fixed pipework • excessive pump speed • 'kettling' of boiler due to lack of a by-pass and/or pump over-run • incorrect positioning of open vent and/or cold feed.
Breakdown of boilers	Boilers with cast iron heat exchangers tend to be more durable than boilers with lightweight heat exchangers. Ignition devices that are used infrequently and thermocouples are two areas particularly prone to breaking down.

Materials

COMMON DEFECTS	CRITERIA/GUIDANCE
System freezes in cold weather	Freezing of the heating system is unlikely in dwellings in constant occupation. However, improved standards of floor and ceiling insulation means that heating pipes in unheated roof or floor voids are now more at risk of freezing. All heating systems should incorporate sufficient draincocks to ensure complete drainage of the system is possible or, alternately, a frost thermostat can be incorporated into the control system. All pipes in locations outside the heated area should be insulated sufficiently. Anti-freeze solutions can be used but can lead to problems with corrosion, leakage and deposition of solids.

HAPM GUIDANCE

If properly installed, it will be assumed that copper pipework will last for the duration of the Major Defects Insurance Policy. Boilers, radiators, cylinders, pumps, valves and controllers will require replacement during the period of the Policy.

Workmanship

COMMON DEFECTS	CRITERIA/GUIDANCE
Inadequate support for pipework	Copper pipework should be clipped as follows: • 15mm pipework at 1.2m centres on horizontal pipework and 1.8m centres on vertical pipework • 22mm and 28mm pipework at 1.8m centres on horizontal pipework and 2.4m centres on vertical pipework. Clips with a band across the front of the pipe hold the pipe more securely than push-fit clips. Double clips should be used when running a pair of pipes together.
Burn damage to adjoining surfaces	When using blowtorches, the adjoining surfaces should be protected from damage by using a heat-resistant pad.
Capillary joints not wiped	Since the acid in flux can cause long term corrosion and green staining, excess flux should be removed from soldered joints by wiping with a damp cloth.
Inadequate attention to making good	Pipes should be sleeved through walls and floors and a mastic seal applied to the annulus between the sleeve and the pipe. Particular care should be taken when making good the surfaces around and between pipes which is always difficult.
Pipework joints not made well	Pipes should be cut square and pushed fully into fittings. Internal burrs should be removed otherwise they may cause blockages and/or restrict the flow in the pipe.
Untidy pipework	Pipework should be as neat and as unobtrusive as possible. Where pipes are required to cross, special crossover fittings should be used. Pipework layouts should be consistent between dwellings on the same site.
Pulled bends not made well	Pulled bends should be used only for offsets. Any kinking or rippling of the pipe wall should not be accepted. Bends should be made with a pipe bender rather than springs.
Use of taper thread fittings to connect to male bosses on hot water cylinders	To connect male bosses on hot water cylinders, parallel thread fittings should be used in preference to taper thread fittings. A 'cone joint' and back-nut should also be used to make maintenance easier.
Incorrect fitting of circulation pumps	Pumps should always be fixed securely to the wall using purpose-made brackets. Pumps should never be allowed to 'hang' on pipework. The best position for the pump is with the inlet and outlet vertical so that sludge does not build up within the pump body.
Pipe layout induces reverse circulation when heating is switched off	Reverse circulation when the heating is switched off can usually be avoided by 'commoning' the heating circuit returns before linking them to the primary return to the boiler.

Workmanship

COMMON DEFECTS	CRITERIA/GUIDANCE
Two or more pipes run in single notch in joist	Having two or more pipes running in a single notch in a joist is bad practice for a variety of reasons. Only one pipe should be placed in each notch. The pipe should be placed on a felt cushion and have a steel protective plate fitted over .

HAPM GUIDANCE

A high standard of workmanship should be demanded by site supervision to reflect the long-term duration of the Major Defects Insurance Policy. The Technical Audit Unit Site Inspectors will undertake a 'quality count'.

Design

COMMON DEFECTS	CRITERIA/GUIDANCE
Blockages and air locking within system caused by using single-stack system beyond design limits	BS 5572: 1978 *Code of practice for sanitary pipework* provides guidance on the design and construction of above ground drainage. Design rules for single-stack drainage systems are included in *Approved Document H.* Generally, a single-stack system can be used for not more than two groups of appliances per floor.
Pipework obstruction of ventilators, windows, doors or placement of fixtures and fittings	Careful design in the initial stages will eliminate the need for awkward and unsightly routing of pipework.
No rodding facilities at the base of stacks and rainwater downpipes	The provision of rodding eyes to all stacks and downpipes is inexpensive and will make maintenance of the drainage system easier.
Improper placement of soil stack terminals	Soil stack terminals should be 900mm above any opening window or other opening within a 3 metre radius (including rooflights and dormers).
No provision for washing machine branches	Washing machines and dishwashers should be anticipated and provision for them included in the design of the above ground drainage system.
No consideration for thermal expansion	'O' ring systems are naturally able to cope with thermal movement. Solvent weld systems require movement joints in long straight runs of pipe.
Surface water discharging into a foul water sewer	Surface water should not be discharged into a foul water sewer where a separate sewer for surface water is provided.
Inadequate support for gutters	Support brackets for the gutters should be at a maximum spacing of 1000mm, or 600mm where the gutter is inaccessible. There should be additional brackets at angles and outlets.
Gutters not strong enough	Gutters and their fixings should be able to cope with snow loads and ladder loads.
Insufficient size and spacing of rainwater downpipes	The number of rainwater downpipes should be calculated by reference to BS 6367: 1983 *Code of practice for drainage of roofs and paved areas*, taking into account factors such as: • the fall of the gutter • the positioning of the downpipe • the flow resistance offered by corners. Concealed flat roof areas should be drained by two or more outlets, or provided with an overflow in a conspicuous position.

HAPM GUIDANCE

Designers should confirm that the above ground drainage system complies with *Approved Document H.* Pipes and fittings should be designed or selected from appropriate proprietary components using manufacturers' design guidance where it is directly relevant. Scheme drawings should indicate clearly the positions of downpipes, vent stacks and horizontal pipe runs.

Details

COMMON DEFECTS	CRITERIA/GUIDANCE
Inaccessibility of pipes for maintenance in cupboards, ductwork or kitchen units	BS 8313: 1989 *Code of practice for accommodation of building services in ducts* gives guidance.
Inaccessibility of traps under baths and showers	Designers should consider the access required for maintenance for baths and showers, and detail accordingly.
Breaking of seals on traps	Where there is a risk of siphonage, ventilation pipework should be provided to all appliances.
Dpm perforated by base of stack	Significant problems will result where there is a ground water problem, and/or in basement situations. Dpms should be turned up around perforating pipes and sealed using adhesive tape.
WC pans leak at stack because set on 'flexible' floor deck	There is a particular problem with WC pans leaking when set on 'floating' floors. Connecting the WC via a branch and using flexible joints may ease the problem.
Gutter-type rise and fall brackets lever up bricks above	To prevent levering up of the bricks, at least two courses of brickwork are required above the position of the brackets.
Differential movement between rainwater outlet in flat roof and flexible deck	Differential movement can occur between the rainwater outlet and the flexible deck due to the deflection and shrinkage of the supporting timber structure, or expansion of the rainwater pipe resulting in the rainwater outlet ending up as the high point in a flat roof. The rainwater outlet or pipe should be detailed to accommodate such movement.
PVC rainwater downpipes or waste pipes not properly connected to clayware gullies or pipes	Special proprietary connectors should always be used. (Using a mortar fillet supported on a screwed-up cement bag is not acceptable!)

HAPM GUIDANCE

Gutter details should be included with roof details. A schematic diagram of the soil stack should contain adequate information for general assessment. Details may be necessary to show pipe enclosures, and the method of passing pipes through walls of enclosure.

Materials

COMMON DEFECTS	CRITERIA/GUIDANCE
No overflow provision	All components should be selected to give sufficient overflow provision. Note that some sink designs do not make such provision.
Air admittance valves prone to freezing and/or tampering	Only air admittance valves possessing a BBA Agrément Certificate, plus an anti-condensation feature, should be considered. Accessibility for maintenance should be considered but publicly-accessible locations should be avoided.
Selection of unsuitable materials	The material should be adequate for its intended purpose, ie uPVC should not be used where disposal of chemical cleaners, etc is anticipated. As an alternative, ABS(1) plastics (which have a higher softening point than a uPVC) can be considered.
Failure of pipework due to degradability in ultra-violet light where it emerges through external wall above ground	Polypropylene pipework to BS 5254: 1976 *Specification for polypropylene waste pipe and fittings (external diameter 34.6mm, 41.0mm and 54.1mm)* and BS 5255: 1989 *Specification for thermoplastics waste pipe and fittings* is known to degrade on exposure to ultra-violet light and should not be used.
Fixtures and fittings not to British Standard quality	Many non-British Standard items of inferior quality are available but should not be used. Any items not covered by a British Standard should have an appropriate BBA Agrément Certificate.
Design of rainwater system allows pipework to drop apart, especially at gutter connection with downpipes	Only good quality, well-designed rainwater goods should be used. Downpipe brackets should provide positive support to the pipework at fittings. Long offsets should be supported.

HAPM GUIDANCE

Specifiers will not be required to adhere to specific durability criteria for components which are relatively easy to maintain (eg fascia-mounted external gutters). Inaccessible components, such as internal soil and vent stack and internal rainwater stacks, should have a target life in excess of the duration of the Major Defects Insurance Policy.

Workmanship

COMMON DEFECTS	CRITERIA/GUIDANCE
Inadequate support to pipework	BRE *Control of quality in design and construction of dwellings, Volume 3: An aid to site inspection* includes useful checklists.
No provision for pipe expansion	Expansion in plastics pipework should be accommodated by the provision of flexible joints or changes of direction at maximum 1.8m centres.
Inadequate restraint of vertical pipe with flexible connections under or at sockets	Flexible-jointed, vertical stack pipes should be held rigidly at each socket otherwise the whole system may 'settle', resulting in backfalls to branches and stressing from subsequent thermal movement.
Insecure attachment of sinks and washbasins	Sinks and washbasins should be attached firmly in accordance with manufacturers' instructions. Where sinks and washbasins are fixed to lightweight blocks or studwork, extra grounds or noggins should be provided.
No sleeving of pipework through walls and floors	Sleeves should be provided to pipework through walls and floors. The annular space should be suitably sealed to meet any relevant fire or sound insulation requirements.
Gutter expansion/contraction prevented by incorrect sizing of gutter lengths	Care should be taken when selecting gutter lengths. Note that gutter brackets usually indicate the length of engagement necessary for movement to occur safely.
Acute bends and horizontal runs	Above ground drainage systems should be planned carefully since the risk of a blockage occurring is increased with the number of bends and branches in the pipework.
Incorrect dressing of roof felt	Roof felt should be extended into a gutter to prevent water being blown into the eaves.
Gutters fixed before fascia painted	Fascias should be painted before the gutters are fixed.
Inadequate fixing of gutters	Screws should be long enough to fix the gutters securely to the fascia. Since some cellular core fascia board materials will not take screw fixings, a timber block should be provided behind the fascia to take the screw fixing for the gutter bracket.
Failure of internal rainwater pipes	Internal rainwater pipes should be tested to withstand a constant air test pressure of 38mm water gauge for a period of 3 minutes.
Corrosion of fixings	Preferably, stainless steel screws should be used for external fixings since even plated screws will rust where the plating is damaged by the screwdriver. Black japanned screws should not be used.

HAPM GUIDANCE

Builders and Supervisors should pay particular attention to these and similar common problems. The Technical Audit Unit Site Inspectors will undertake a 'quality count'.

Design

COMMON DEFECTS	CRITERIA/GUIDANCE
Overcrowding of services within ductwork due to undersizing of duct, or access to ducts is inadequate or lacking	Many designers treat the ductwork as an afterthought and try to fit service runs in after the layouts are complete. Sufficient ductwork should be designed in at an early stage to provide for all the required services and with enough space to assist in future maintenance.
Non-uniform shape or change of direction in ductwork	To avoid maintenance problems, ductwork runs should be kept simple and uniform.
Ducts routed through escape routes or lift shafts	Preferably, ducts should not be routed through escape routes or lift shafts. Where possible, consideration should be given to alternative routing. Refer to Clause 5.8 in BS 5655: Part 1: 1986 *Lifts and service lifts: Safety rules for the construction and installation of electric lifts* for guidance.
Bin chute diameters undersized and inadequately ventilated	There are clear guidelines for chute design, including materials, set out in BS 1703: 1977 *Specification for refuse chutes and hoppers*. Chutes should be located separate from, and not adjacent to, a habitable room.
Noise from chutes at unacceptable level	The density of walls separating a refuse chute from a habitable room is controlled by *Approved Document E*.
Confusion between ducts, protected shafts and casings	Services may be accommodated in what, by definition, is either a casing or a protected shaft. The fire resistance requirements for each of these are distinctly different. The construction of a casing is described in Diagram F2 in *Approved Document B2/3/4*. A protected shaft is an element of structure and so requires a specified level of fire resistance both within and outside. Subdivision of a protected shaft to coincide with the junctions with other elements is not necessary.
Accommodation of gas and electric services in a common duct	For flats and maisonettes over two storeys high, a common duct for gas and electric services is not permitted. Refer to CP 3: Chapter IV: Part 1: 1971 *Code of basic data for the design of buildings: Precautions against fire: Flats and maisonettes (in blocks over two storeys).*
Inadequate provision for ventilation ducts containing gas mains or services	Where possible, gas pipework should not be enclosed in ducts but when this becomes necessary, the ducts should be ventilated adequately to the outside air at high and low levels at each storey. British Gas should be consulted for the design requirements for gas pipework at an early stage. Refer to the British Gas publication *Gas in Housing*.

HAPM GUIDANCE

Ductwork should be designed in accordance with BS 8313: 1989 *Code of practice for accommodation of building services in ducts*. Bin chutes should comply with BS 1703: 1977.

Details

COMMON DEFECTS	CRITERIA/GUIDANCE
No ventilation to ducts with pipework containing town gas, LPG or natural gas	Where fire stopping subdivides a duct, ventilation will be required to each subdivision. Refer to *The Gas Safety Regulations 1972* and *The Gas Safety (Installation and Use) Regulations 1984*.
Insufficient space in duct to allow for inspection and repair, and no space for additions and alterations to services within ductwork	Careful consideration should be given to the spacing of services within ductwork and also to the degree of demolition required to the duct (together with disruption of other services) to accommodate maintenance and modification. Ideally, between 10% and 15% extra space should normally be allowed in ductwork for later additions or alterations.
Flue pipes in common ducts leading to heat build-up	Where possible, flues should not be run in common ducts but when this becomes necessary, the flue should be insulated. Where the surface temperature of the casing may fall below the dew point of the air in a duct, precautions to prevent surface condensation will be required.
Condensation due to lack of insulation to air ducts	Where extreme changes of temperature in the duct are likely, insulation should be provided.
Overheating of power cable in vertical ducts	To prevent overheating of the power cable in vertical ducts, internal barriers should be provided in accordance with the *IEE Regulations*. Alternatively, enough ventilation to prevent overheating should be provided.
Ductwork exposed due to insufficient ceiling space	Where possible, to avoid the need for exposed ducts, services should be accommodated in ceiling voids or dropped ceilings.
Invasion of ducts by vermin such as rodents and pharaoh ants	Care should be taken to ensure that the materials used in ducts should not encourage vermin, and that potential entry points are sealed.
Insufficient resistance to fire	As part of the detail design for the overall fire engineering requirements for the building, the following should be assessed for all ducts: • combustibility • fire resistance • resistance to fire spread. For guidance on ventilation and air conditioning ducts, refer to BS 476: Part 24: 1987 *Fire tests on building materials and structures: Method for determination of the fire resistance of ventilation ducts* and BS 5588: Part 9: 1989 *Fire precautions in the design and construction of buildings: Code of practice for ventilation and air conditioning ductwork.*
Floor ducts covered by tiling or vinyl sheet	Floor ducts should have screw down, removable panels. Adhesive-fixed floor finishes should not be continuous over a duct.
Sound insulation of floor compromised by floor ducts	Floor ducts should not be considered in situations where they could reduce the sound-insulating properties of a separating floor.

HAPM GUIDANCE

Locations of ducts should be clear on drawings. Horizontal sections through ducts and shafts should show the positions of all services to demonstrate adequate space. A section through a vertical duct or shaft at floor level should show the method of fire stopping, if required.

Materials

COMMON DEFECTS	CRITERIA/GUIDANCE
Corrosion of metal items in ducts	Ducts may contain many metal brackets and associated fixings. Electrical cabling may be carried on steel cable trays. To prevent corrosion, all such metalwork should be hot-dip galvanized.

HAPM GUIDANCE

All materials for ducts, shafts and chutes should conform to the requirements of the relevant British Standards and the requirements of the *Building Regulations* and the *Fire Precaution Act*.

Workmanship

COMMON DEFECTS	CRITERIA/GUIDANCE
Omission of fire stopping and building rubbish left in ducts and shafts	Before closing the duct, checks should be carried out to ensure that fire stopping is complete and all rubbish has been cleared.
Misalignment of bin chute liners giving rise to damage and blockage	Chute liner sections should be supported adequately and aligned during each stage of construction since faults are very difficult to rectify at a later stage.
Omission of access panels or access panels difficult to open	Preferably, access should be provided at all fittings requiring maintenance and should be easily operable with the limited use of simple tools.
Omission of sufficient supporting structure to ductwork panels	The supporting structure to the ductwork panels should be checked before it is covered.
Condensation on pipes within ducts	Cold water pipes and air ducts should be insulated in accordance with BS 5422: 1977 *Specification for the use of thermal insulating materials* and BS 5970: 1981 *Code of practice for thermal insulation of pipework and equipment (in the temperature range -100°C to +870°C).*
Uneven floor duct covers	Care should be taken to align accurately the covers to floor ducts with the floor surfaces and any neighbouring covers.
Horizontal pipe ducts in screeds not fully bedded	Duct covers to screed ducts usually transmit live loads (ie furniture and foot traffic) to the sides and bottom of the duct casing. When the duct is not fully bedded, or the screed not well-compacted against the sides of the duct casing, displacement of the duct may occur.

HAPM GUIDANCE

Builders and Supervisors should pay particular attention to these and similar common problems. The Technical Audit Unit Site Inspectors will undertake a 'quality count' and a poor result will be reported for an assessment of risk.

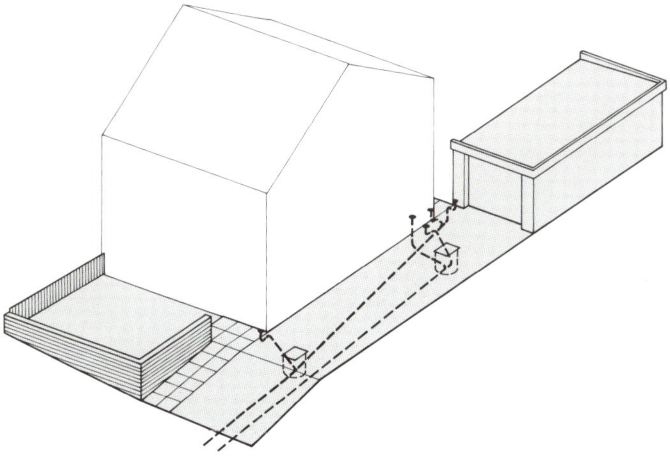

HAPM

HOUSING ASSOCIATION PROPERTY MUTUAL LIMITED

BELOW GROUND DRAINAGE

EXTERNAL WORKS

OUTBUILDINGS

7

Design

COMMON DEFECTS	CRITERIA/GUIDANCE
Site conditions not fully investigated	Existing sewerage systems should be available to accept the proposed discharges. The following should be established (taking surcharge conditions into account): • water table levels • flood water levels • levels at which connections may be made into existing sewers.
Failure of drainage system due to waterlogging	Field drains may have to be provided in the surrounding area. To avoid waterlogging on sloping sites, a one metre wide flat area sloping slightly away from the dwelling and at least 150mm below the level of the dpc should be provided. Soil should be kept below the level of the airbricks. Where the water table level is above the level of the sewer invert, particular care should be taken in the construction of all pipe joints and manholes to ensure they are watertight.
Insufficient access to pipes for maintenance	Each dwelling should have its own manhole or inspection chamber within the curtilage of the property. Access for rodding should be provided at all bends and junctions. The spacing of access points and rodding eyes is clearly specified in BS 8301: 1985 *Code of practice for building drainage*. Although there may be technical and economic advantages in providing a drainage system to serve more than one property, such an arrangement can cause difficulties in conveyancing or the apportionment of maintenance costs.
Gradients too steep or too shallow	The size of the pipe and the gradient at which it is to be laid should be such that the risk of induced trap siphonage at peak flow is minimised by ensuring sufficient air movement in the drain. Provided that at least one WC is connected, the choice of gradient should be sufficient to maintain a self-cleansing velocity, as follows: • for a peak flow of 1 litre/sec, a 100mm diameter pipe should not be laid flatter than 1 : 40 • for a peak flow greater than 1 litre/sec, a 100mm diameter pipe should not be laid flatter than 1 : 80. High velocities of sewage flow arising from steep gradients can cause erosion of pipes or deposition of solids. In situations where a very steep gradient would result, a back-drop manhole should be considered.
Pipes laid at shallow depths	Where pipes have to be laid at less than 900mm below the wearing surface of roads, or less than 600mm below the surface of fields and gardens, protection should be provided against: • site construction traffic • future building works • agricultural activities • erection of fences • frost, etc. This protection will normally consist of a concrete surround to the pipe, or concrete bridging slabs.

Design

COMMON DEFECTS	CRITERIA/GUIDANCE
Manholes located in soft ground	Where possible, manholes should be located in hard paved areas.

HAPM GUIDANCE

The drainage layout showing pipe sizes and gradients should be designed to conform to the requirements of *Approved Document H* and to BS 8301: 1985 and to any local legislation that may be in force. Drainage manufacturers' literature forms a useful guide to design and those which are particularly clear are Hepworth's *Technical Handbook* and Naylor Bros Architects' *Drainage Design Handbook*. Negotiations should be made at an early stage with the local authority, and a Section 18 Agreement completed, with a view to the works being adopted by the local authority on completion. Refer also to the Water Authorities Association publication *Sewers for Adoption – a design and construction guide for developers*.

Details

COMMON DEFECTS	CRITERIA/GUIDANCE
Failure of joints	In mining areas and in areas on shrinkable clays, particularly where there are trees or trees have been removed, flexible joints should be used to accommodate subsidence or heave. Where a rigid system is adopted, the cement/sand joints should not be stronger than 1 : 2. All proprietary joints should be formed strictly in accordance with the manufacturer's instructions.
Failure of pipes	At the points of entry into buildings or manholes, lintels or sleeves should be provided to accommodate differential settlement between the pipe and the building.
Inadequacy of soakaways	Soakaways should be considered only when no alternative drainage system is available. Where soakaways are to be used, they should be constructed only in porous ground, or on suitable land sloping away from a building. All soakaways should be at least 5 metres away from the foundations of a building. Refer to BRE *Digest 151* for details of the test for measuring the rate of water percolation into the surrounding ground. In sandy or silty ground, soakaways tend to clog up through retention of the fine material. To help prevent this, the top of the soakaway should be protected by a layer of woven nylon fabric to exclude fine materials.
Instability of pipes	Where pipes are laid through made ground, and in cases where foul and surface water pipes are in a common trench, careful attention should be given to details, such as the requirement for concrete surrounds, especially to the shallower pipe.
Leaking manholes, and manholes of insufficient size	In waterlogged ground, and where the water table is above the base of the manhole, joints should be made watertight by using a mastic sealant or a rubber ring joint. Where the depth of the manhole or inspection chamber is greater than one metre, the inside dimensions should be increased to allow access for working space. Where the manhole or inspection chamber is in unstable ground, or subject to exceptional loads, it should be provided with a 150mm thick concrete surround, with care being taken to compact this concrete under and around the incoming and outgoing pipes.
Blockage of or damage to pipes by entry of tree roots	Where pipes cannot be laid below the level of tree roots, a concrete surround should be provided to the pipes.
Instability under buildings	Runs of pipes beneath buildings should be detailed so that there is little possibility of failure, bearing in mind their future inaccessibility. Provision should be made for any differential settlement at the point where a drain emerges from under a building. Whenever practicable, drains should be laid outside buildings.

HAPM GUIDANCE

Designs should include typical cross-sections through drains, chambers and gullies to avoid doubt on site as to how to construct the system.

Materials

COMMON DEFECTS	CRITERIA/GUIDANCE
Failure of pipework due to mixture of pipe materials	To reduce the risk of failure of the pipework, only one type of pipe should be used in any one run of pipes between manholes.
Softening or distortion of pipes and fittings	Only uPVC products should be used. Other types of plastics, such as polypropylene, tend to disintegrate under ultra-violet photo-oxidation. Pitch fibre pipework should not be used.
Deterioration of pipes and manholes due to chemical action	The physical and chemical nature of the ground should be investigated, particularly for the presence of sulphates, so that components can be selected which have resistance to such attack. Engineering-class bricks in manholes usually provide sufficient resistance to chemical attack. Generally, sulphate-resisting cement should be used for concrete pipes and for mortar in manhole brickwork. Super-sulphated cements should not be used.
Failure of manhole covers and slabs	Manhole covers and slabs should be designed to carry the likely traffic but cases of failure continue to arise due to traffic overload. Refer to BS 5400: Part 2: 1978 *Steel, concrete and composite bridges: Specification for loads* for the specifications for traffic loads. Manhole cover slabs are usually precast concrete and should conform to BS 5911: Part 2: 1982 *Precast concrete pipes, fittings and ancillary products: Specification for inspection chambers and street gullies.* Manhole covers may be cast iron or cast steel with the grade selected to suit the site conditions and should conform to BS 497: Part 1: 1976 *Specification for manhole covers, road gully gratings and frames for drainage purposes: Cast iron and cast steel.*
Inadequate attention to bedding and back filling materials	Generally, granular material is used for pipe bedding. Suitable materials include: • single-sized aggregate of 10mm, 14mm or 20mm • graded aggregate 20mm to 5mm. Selected material from trench excavation is suitable for back filling over rigid pipes and flexible pipes provided it is readily compactible. Flexible pipes should first be covered with a 100mm layer of bedding material before back filling. Refer to BS 8301: 1985 *Code of practice for building drainage* for specific guidance.

HAPM GUIDANCE

All materials for pipes, pipe joints, bedding, surrounds, and manholes should conform to the requirements of the relevant British Standards. Types of brick, cement and mix to be used should be selected to ensure satisfactory service for the life of the drainage system which has to equate to the life of the property being drained.

Since British Standards cover all generally used drainage materials, any materials not so covered should be supported, if possible, by independent test reports.

Workmanship

COMMON DEFECTS	CRITERIA/GUIDANCE
Damage to existing drains and services	Records of existing pipes and services should be held on site and excavation should always proceed cautiously. Records should be kept of all pipes and services that are encountered. Any defect found in any pipe or service should be reported to the appropriate authority.
Errors in setting out	Runs of pipes between manholes should be straight and laid to an even gradient. The line should be checked before excavation for possible obstructions, such as manholes, foundations, lamp posts, etc. The setting out of manholes should be checked to ensure the manholes are fully located in hard paved areas. The *Manual of Setting-out Procedure* published by the Construction Industry Research and Information Association (CIRIA) gives good guidance.
Collapse of excavations	Excavations should not be started until all the material required is on site, and they should be kept open for as short a time as possible. Sufficient support for the sides of excavations should be provided. Pumping equipment should be available. Excavations should be back filled as soon as tests for the watertightness of joints are satisfactory. CIRIA Report No 97 *Trenching Practice* gives good guidance.
Soft spots in excavations	Soft spots in excavations should be dug out and the resulting void filled with granular material or lean-mix concrete. Similar treatment should be given to any trench bottoms that have been allowed to become soft. The bedding for each pipe should be as uniform as possible throughout.
Excessive excavation of trenches	Excessive excavation of the depth of trenches should be made good with suitable compacted granular material or lean mix concrete. Excessive excavation of the width of trenches should be avoided as it can lead to 'snaking' of drain runs laid in granular material.
Excavation near existing foundations	Excavations within one metre (laterally) of existing foundations should be back filled with lean-mix concrete to the level of the foundations. Excavations over one metre from a foundation should be back filled with lean-mix concrete up to a 45° cut-off line.
Inadequate support for pipes	Where the pipework is inadequately supported, fracture can occur. The pipe should be in contact with its bed for the whole length of the barrel. The bedding material should be cut away locally to accommodate sockets in pipe runs.
Leaking joints	Proprietary joints should be installed strictly in accordance with the manufacturer's instructions. The mortar for cement/sand joints should not be stronger than 1 : 2 and should be placed around the full periphery of the joint. The joints in each run of pipes between manholes should be tested on completion.

BELOW GROUND DRAINAGE

Workmanship

COMMON DEFECTS	CRITERIA/GUIDANCE
Blockage of pipes	Most blockages occur initially due to builder's rubbish not being removed. Therefore, a careful final check of all pipework is required.
Damage to pipes during back filling	Back fill material should be placed equally on each side of the pipe and should be fully consolidated by ramming, in layers of about 150mm depth. The back fill material should be free from large stones.
Defects in flexible joints	Where flexible joints are required in pipe runs, the same flexibility should be maintained through the full cross-section of any concrete surround by providing compressible material at the face of each pipe socket or sleeve.
Defects in manholes	Manholes should be checked to ensure the joints are watertight. Any leakage in a brick manhole would normally be through joints in the brickwork. Therefore, as the bricklaying proceeds, attention should be paid to: • bedding all bricks fully • filling all joints • flush pointing.
Poor joints at channel connections and between benching and channels in manholes	Where side branches discharge into main channels in the manholes, this should be in the direction of flow. Vertical and side benching should be shaped to contain the flow without permitting fouling, and to accommodate convenient rodding of branch drains. The channels should be bedded and jointed in 1 : 3, cement : sand, mortar and benching floated to a hard smooth surface with a coat of 1 : 3, cement : sand, mortar laid monolithic with the benching in a single operation.
Inadequate support to gullies, rest-bends, etc	All features at ground level should be supported securely – generally by a concrete surround.

HAPM GUIDANCE

Site inspections and tests are necessary before final handing over to ensure that materials are installed properly and that the system functions as intended without creating any health risk or nuisance. All tests should be carried out and recorded in the presence of an authorised testing officer and in conjunction with the local authority.

Generally, water tests are preferred since they relate to conditions found in practice. On completion of testing, the testing officer should observe the flow of water out of the system to ensure that the system was completely filled. The drainage system has to be constructed such that it will last at least for the life of the building. On completion of all the works of a project, the complete drainage installation should then be inspected and checked under normal working conditions.

Design

COMMON DEFECTS	CRITERIA/GUIDANCE
Inadequate drainage to paths and drives	Drainage should always be away from the building. All impervious surfaces should drain either onto permeable garden areas or into trapped gullies connected to soakaways or the drainage system.
Paths and driveways too narrow	Pedestrian ways should be a minimum of 600mm wide. Vehicle access ways (for private cars) should be a minimum of 2.4m wide but this should be increased at sharp curves and corners.
Inadequate external lighting for security and safety	Durable and vandalproof lighting should be provided for hazards, such as steps and ramps. Dark recesses where people may 'lurk' should also be lit. A reasonable level of illumination should be provided to unadopted paths, pavings and parking.
Stability of embankments impaired by landscaping schemes	Landscaping should be designed so that the finished slope of the ground is not steeper than the angle of repose of the material, less a margin of -5°.
No movement joints provided in boundary walls	To accommodate cyclical thermal movements, 10mm wide movement joints should be provided: • at about 12m centres for clay bricks • at about 8m centres for calcium silicate bricks • at about 6m centres for concrete blocks. Where movement joints are provided, the stability of the wall should also be considered, for example for 100mm thick walls, a pier would be required at the end of each panel.
Impaired stability of openings in boundary walls	Piers should be provided adjacent to all openings in boundary walls.
Insufficient resistance of boundary walls to wind	Boundary walls should be designed to suit the wind loads specified in CP 3: Chapter V: Part 2: 1972 *Code of basic data for the design of buildings: Loading: Wind loads.* The mortar should be specified to suit the strength required. Refer to BDA *Design of Freestanding Walls.*
Design problems due to effect of trees	Soil shrinkage caused by the removal of water from the ground by live trees can result in subsidence. Soil swelling by the recovery of moisture in the ground following tree removal can result in heave. These problems are particularly severe in shrinkable clay soils. Both conditions can disturb significantly paths, pavings, planters, garden walls etc. In time, some reconstruction may be inevitable, unless a very robust design is adopted.
No pedestrian guard rails or vehicle barriers provided	Where there is drop of 600mm or more beside any path or driveway, guarding should be provided. Refer to BS 6180: 1982 *Code of practice for protective barriers in and about buildings* for guidance.

Design

COMMON DEFECTS	CRITERIA/GUIDANCE
Paths and drives too steep	Landscaping should be carried out so that: • the gradient of any driveway is not greater than 1 : 20 (5%) • the gradient of any path is not greater than 1 : 12 (8.33%). Where the path cannot be constructed to this gradient, steps should be provided.
Inadequate specification for bases and sub-bases	Bases and sub-bases should be specified in terms of: • suitable materials to be used • compacted thicknesses required • weight of roller to be used • number of passes to be made by the roller. *Road Note 29* is a good guide to the requirements. Refer also to *Information Sheet 1* published by the British Aggregate Construction Materials Industries.

HAPM GUIDANCE

All external works should be designed to conform to any requirements of the *Building Regulations*, relevant British Standards and the local authority. All agreements made with the local authority regarding the adoption of roads, footpaths, requirements for vehicle cross-overs over footpaths, driveways and access paths adjacent to public highways should be recorded.

Details

COMMON DEFECTS	CRITERIA/GUIDANCE
Build-up of ground water allowed by retaining walls	The drainage layer behind the wall should be gravel of about 40mm single size or crushed rock, and should be 450-600mm thick. The drainage layer should drain into a semi-porous drain of 100-150mm diameter, which should lead into the surface water drainage system or to a soakaway. Alternatively, weepholes through the wall may be acceptable.
Moisture penetration through retaining walls	The rear faces of retaining walls should be suitably tanked. The sophistication of the tanking should reflect the ground water conditions and the characteristics of the bricks or concrete forming the wall. Note that a sand and cement render is usually sufficient.
Stability of freestanding walls impaired by membrane-type dpc	Where a dpc is required, it should be provided by using engineering-quality bricks in 1 : 3 mortar to 150mm above the level of the ground.
Inadequate weather protection to freestanding walls	When flush engineering brick cappings are used, the walls may not receive the required weather protection. Therefore, freestanding walls should be protected by copings which: • throw water away from the face of the wall • stop water getting into the top of the wall • overhang the face of the wall by 45-50mm • provide a drip at least 25mm from the face of the wall. See Topic 7.2, Materials, for more guidance.
No path edging provided	Rapid failure of flexible pavings may occur if sufficient perimeter restraint is not provided.
Concrete kerbs or edgings out of line	Concrete kerbs and edgings should be fully bedded and haunched with concrete not weaker than Grade C20. The sub-base should extend under and beyond the kerbs, unless the foundation provided is sufficient to support the kerb and haunching. Refer to BS 7263: Part 2: 1990 *Precast concrete flags, kerbs, channels and quadrants: Code of practice for laying* for guidance.
Rot in timber fence posts at ground level	Timber fence posts should always be treated with preservative applied by pressure-impregnation. Posts should be set on a shingle layer to allow drainage and should be surrounded with well-rammed excavated material or a very weak concrete mix.
No protection provided to service pipes or cables	Service or drain pipes and cables in footpaths or driveways should be more than 600mm deep. Where this is not possible, they should be protected with a concrete surround.
No movement joints in paths or driveways, and edges not sealed	To accommodate cyclical thermal movements in paths and driveways, movement joints should be provided at approximately 4 metre centres, and at corners. Edge joints in concrete surfaces abutting buildings should be filled with a joint filler and sealed.

Details

COMMON DEFECTS	CRITERIA/GUIDANCE
Steps awkward to use	Steps should be no steeper than permitted by *Approved Document K*, but in practice, and for convenience, they are usually made more shallow. Steps should have a small fall to allow drainage.

HAPM GUIDANCE

Attention should be given to details and finishes ensuring compatibility of all materials and having regard to the conditions to which they will be subject to resist wear and deterioration during the life of the dwellings. Sufficiently detailed construction information should be provided for all external works features.

Materials

COMMON DEFECTS	CRITERIA/GUIDANCE
Rot in timber edge boards and pegs to footpath edges and timber fences	All timber should be treated with a fungicidal preservative applied by pressure-impregnation. Treatment cycles should be specified to suit the type of timber used.
Insufficient resistance of concrete paths and driveways to frost	The top 50mm of all exposed concrete surfaces should be air-entrained to give an air-entrainment of about 4.5%. The air-entraining agent should include a water reducing agent and should be added to the concrete mix strictly in accordance with the manufacturer's instructions. The air-entraining agent should conform to the requirements of BS 5075: Part 1: 1982 *Concrete admixtures: Specification for accelerating admixtures, retarding admixtures and water reducing admixtures,* and should not contain any calcium chloride. The concrete should be at least C25 quality (25N/mm^2 at 28 days) and should contain a minimum cement content of 300kg/m^3.
Deterioration of brick garden walls	Brick quality should be confirmed by the brick manufacturer as suitable for freestanding walls. Brick copings are most vulnerable to deterioration but, where the coping does not provide for water shedding, the brickwork below the coping is also very vulnerable. Fletton brickwork should not be used for garden walls.
Deterioration of bricks or blocks at or below ground level	In walls below the level of the ground, and up to the level of the dpc, one of the following should be used: • engineering-quality bricks designation FL • engineering bricks Class B to BS 3921: 1985 *Specification for clay bricks.* In other situations below ground level, sulphate-resisting bricks, such as calcium silicate, or sulphate-resisting blocks should be used. In all cases, it is advisable to use sulphate-resisting cement for all mortar and concrete below the level of the dpc.
Corrosion of steel gates, guard rails, lamp posts	All mild steel components should be supplied hot-dipped galvanized and given a paint treatment. Note that these components may require repainting at regular intervals, depending on the aggressiveness of the atmosphere in the locality.
Garden gates, guard rails, lamp posts, street furniture not vandalproof	All external items adjacent to public areas should be of a robust design and, as far as possible, vandalproof.
Paths and driveways slippery when wet	Concrete should have a combed or textured finish and should be set to a fall that will shed water but will still provide a foothold when wet. Hot asphalt surfaces should be finished with a dimpling roller. Clay or concrete tiles should have a textured finish.

Materials

COMMON DEFECTS	CRITERIA/GUIDANCE
Manhole covers not suited to vehicle loadings	The manufacturer's technical information should be checked to ensure the manhole covers are suitable for the likely vehicle loadings (remembering that removal lorries and refuse vehicles do not always respect normal access restrictions!).

HAPM GUIDANCE

All materials, apart from having to conform to the relevant British Standards or being supported by a BBA Agrément Certificate, should be robust and capable of resisting hard wear and abuse.

Workmanship

COMMON DEFECTS	CRITERIA/GUIDANCE
Fill not compacted in layers	Back filling of excavations should be completed in layers not exceeding 225mm of loose depth. Each layer of fill should be fully compacted before the next layer is placed.
Backfalls on hard surfaces	Paths and driveways should be constructed to falls so that water drains away from buildings and into gullies or onto pervious areas reducing the risk of ice formation in cold weather. Note that surface water should not drain onto public highways.
Fence posts not set deep enough	Generally, fence posts should be embedded in the ground for a depth equivalent to one-third of the exposed height of the fence. For fences in exposed areas, the advice of a Structural Engineer should be sought.
Builder's rubbish buried under garden areas	Builder's waste, rubbish etc occurring within 400mm of the finished ground level should be cleared from the site. Garden levels should be reinstated to uniform levels appropriate to the level of the building, adjacent roads and other properties. A minimum of 200mm of topsoil should be provided over all garden areas.
Inadequate protection to new concrete	Newly-poured concrete should be protected: • in hot weather against too rapid drying out to prevent shrinkage cracks • in cold weather against frost damage or freezing. Proprietary curing compounds or membranes may be used.
Poor compaction of sub-base and pavement to paths, pavements, roads and parking areas	The sub-soil should be non-organic, levelled and well-consolidated, and free from mud and slurry. The durability of a finished path will depend heavily on how well the various layers are compacted. The design information should specify the requirements for compaction in terms of the weight of the roller to be used and the number of passes to be made by the roller, which should then be monitored on site during construction.

HAPM GUIDANCE

External works are usually the last items to be completed and are often rushed to complete the contract on time. The work area may have been a depository for spare and discarded materials and the works often lack supervision. As a result, there are many premature failures. The Technical Audit Unit Site Inspectors will include external works in the 'quality count'.

Design

COMMON DEFECTS	CRITERIA/GUIDANCE
Cracking away of garages attached to houses	The footings for garages should be of the same type as those for the main dwelling, and preferably should be taken to the same depth. Due to the potential for differential footing loading, it is often advisable to form a movement joint at the junction of the garage and main dwelling.
Garage walls unstable	Masonry should be designed to BS 5628: Part 1: 1978 *Code of practice for use of masonry: Structural use of unreinforced masonry.* Normally, piers are required in the long walls running from the front to the rear of garages.
No movement joints provided between garages in multi-unit blocks	A movement joint should be provided between each pair of garages in a multi-unit block.
Flat roofs allow water ponding	The size of timber joists may be selected from *Approved Document A1/2.* The fall of the roof should be designed as 1 : 40 so that after deflection of the joists has taken place, the final fall may approximate to 1 : 80.
Cracking and breaking up of floors of outbuildings	Where an outbuilding is not attached to a dwelling, it is usually more economical to build it on a shallow raft-type foundation. The sub-soil should be consolidated and a layer of compacted hardcore provided before the concrete raft foundation is laid. See Topic 7.3, Details, for more guidance.
Rainwater not drained away from outbuildings	Gutters and downpipes connected to the surface water drainage system or to soakaways should be provided on each building.
Water allowed to drain into garages or bin stores	Where garages or bin stores are situated at the lower ends of slopes, drainage channels should be located across their fronts to prevent water draining into them. The floors should then be constructed to fall towards the front with a minimum fall of 1 : 80.

HAPM GUIDANCE

All outbuildings should be designed to conform to any requirements of the *Building Regulations* and relevant British Standards.

Details

COMMON DEFECTS	CRITERIA/GUIDANCE
Floors slippery when wet	The concrete should have a combed or textured finish and should be set to a fall that will shed water but will still provide a foothold when wet.
Roofs lift off in strong winds	Holding down straps should be provided to tie roofs down securely to walls to resist wind uplift.
Omission of restraint straps from walls and roofs	Restraint straps should be provided so that the walls of outbuildings are tied laterally to the roofs to assist in the stability of the walls.
Garages, garden sheds, communal garden sheds not secured properly and not vandalproof	All buildings should be provided with locking devices and should be of a robust construction and, as far as possible, vandalproof.
Cracking and breaking up of floors of outbuildings	The same care and attention should be given to outbuildings as to the main structure. Raft foundations are more common for outbuildings and should be fully detailed at perimeters, and reinforcement clearly indicated. Where deep fill is likely to be used, suspended floors should be considered.
Subsidence or heave near trees	Outbuildings constructed too close to trees can be damaged by subsidence or heave, especially in areas of shrinkable clay soils. Therefore, outbuildings should not be constructed near trees, and trees should not be planted near outbuildings. See Topic 7.2 for more guidance.
Outbuildings suffer from rising damp	Outbuildings should be provided with a dpm below the concrete slab. The dpm should be continuous with the dpcs in the walls, and should have its joints lapped a minimum of 150mm and sealed.
Paths and garden areas located above dpc level	Paths and garden areas should be located at least 150mm below the level of the dpc of any adjacent outbuilding.
Roof perimeters do not overhang and protect walls	Since outbuildings are usually single skin and may be very exposed, the roof perimeters should provide them with a reasonable level of protection.
Damage to outward opening doors by gusting wind	A means of restraining the door leaf should be provided, perhaps with a door closer.

HAPM GUIDANCE

Technical Audit Unit Site Inspectors will pay special attention to outbuildings which may not receive the same care and attention to detail during construction.

Materials

COMMON DEFECTS	CRITERIA/GUIDANCE
Breaking up of exposed or semi-exposed concrete floors in frosty weather	Garage and bin store floors are often exposed so the concrete should contain an air-entraining agent to give an air-entrainment of about 4.5%. The air-entraining agent should include a water reducing agent and should be added to the concrete mix strictly in accordance with the manufacturer's instructions. The air-entraining agent should conform to the requirements of BS 5075: Part 1: 1982 *Concrete admixtures: Specification for accelerating admixtures, retarding admixtures and water reducing admixtures,* and should not contain any calcium chloride.
Deterioration of bricks or blocks below ground level	In the walls of outbuildings below the level of the ground, and up to the level of the dpc, one of the following should be used: • engineering-quality bricks designation FL • engineering bricks Class B to BS 3921: 1985 *Specification for clay bricks.* In some cases, sulphate-resisting bricks, such as calcium silicate, or sulphate-resisting blocks may be used. In all cases, it is advisable to use sulphate-resisting cement for all mortar and concrete below the level of the dpc. In the majority of outbuildings, the problem of brick or block deterioration will not arise if the building is constructed on a concrete raft foundation.
Premature rot in frames	Timber windows, doors and frames tend to be more vulnerable in outbuildings as they are often 'set forward' and door frames are in contact with brickwork below the level of the dpc. Good preservative treatment and high quality joinery adhesives are essential.

HAPM GUIDANCE

All materials, apart from having to conform to the relevant British Standards or being supported by a BBA Agrément Certificate, should be robust and capable of resisting hard wear and abuse.

Workmanship

COMMON DEFECTS	CRITERIA/GUIDANCE
Leaking prefabricated buildings	A check should be made that all components, including all trims and flashings, are available before erection is attempted. The buildings should be erected strictly in accordance with the manufacturer's instructions.
Inadequate compaction of ground and sub-base below foundations	The ground beneath the sub-base of all raft foundations should be fully compacted before any material is placed. The sub-base should also be fully compacted. See Topic 7.3, Details, for more guidance on the compaction required.
Fill not compacted in layers	Where outbuildings are constructed on sloping ground, filling may be required to provide a level surface. The filling should be completed in layers not exceeding 225mm of loose depth. Each layer of fill should be fully compacted before the next layer is placed. The compaction required is similar to that described for sub-soil under Topic 7.3, Details.
Omission or poor fixing of roof holding down straps	The failure of roofs to outbuildings is common in high exposure areas if particular attention is not given to the provision and fixing of holding down straps.

HAPM GUIDANCE

Outbuildings are usually regarded as of minor importance and, with external works, are among the last items to be completed. The construction is often rushed to complete the contract on time, and poorly supervised due to supervisory staff having been transferred to other projects. The number of defects arising from poor workmanship are therefore out of all proportion to their value. The Technical Audit Unit Site Inspectors will include outbuildings in a 'quality count'.